EMBRYONIC MORTALITY IN FARM ANIMALS

CURRENT TOPICS IN VETERINARY MEDICINE AND ANIMAL SCIENCE

EMBRYONIC MORTALITY IN FARM ANIMALS

A seminar in the CEC programme of coordination of research on Livestock Productivity and Management

Sponsored by the Commission of the European Communities, Directorate-General for Agriculture, Coordination of Agricultural Research

Edited by

J.M. Sreenan and M.G. Diskin
The Agricultural Institute, Western Research Centre
Belclare, Tuam, Eire

1986 **MARTINUS NIJHOFF PUBLISHERS**
a member of the KLUWER ACADEMIC PUBLISHERS GROUP
DORDRECHT / BOSTON / LANCASTER
for
THE COMMISSION OF THE EUROPEAN COMMUNITIES

Distributors

for the United States and Canada: Kluwer Academic Publishers, 190 Old Derby Street, Hingham, MA 02043, USA
for the UK and Ireland: Kluwer Academic Publishers, MTP Press Limited, Falcon House, Queen Square, Lancaster LA1 1RN, UK
for all other countries: Kluwer Academic Publishers Group, Distribution Center, P.O. Box 322, 3300 AH Dordrecht, The Netherlands

Library of Congress Catalog Card Number

Main entry under title:

Embryonic mortality in farm animals.

    (Current topics in veterinary medicine and animal science)
    "Sponsored by the Commission of the European Communities, Directorate-General for Agriculture, Coordination of Agricultural Research."
    Bibliography: p.
    1. Livestock--Embryos--Mortality--Congresses.
I. Sreenan, J. M. II. Diskin, M. G. III. Commission of the European Communities. Coordination of Agricultural Research. IV. Series.
SF887.E43 1985   636.089'83   85-21831

ISBN-13: 978-94-010-8732-2   e-ISBN-13: 978-94-009-5038-2
DOI: 10.1007/978-94-009-5038-2
EUR 9738 EN

CONTENTS

VII

PREFACE

Reproductive wastage is a major inefficiency in all livestock production
with embryonic mortality accounting for a major portion of this loss.

Accordingly the Commission of the European Communities encouraged the
organisation of a seminar on embryonic mortality in farm animals which
was held in Brussels on the 11th and 12th of December 1984.

This book contains the text of the papers, discussions and final summary
presented at that Seminar.

As a background to the Seminar, the extent and timing of embryonic loss
was described for farm animals. Particular consideration was then given
to the various mechanisms and signals, both embryonic and uterine in
origin, that are so far known to be involved in the establishment of
pregnancy. Possible causes of embryonic death including physiological,
endocrinological, genetic and immunological components were outlined
and discussed.

The final summary contains general conclusions from the Seminar and
recommendations for future research work on this topic.

J.M. Sreenan
M.G. Diskin   July 1985.

****

THE EXTENT AND TIMING OF EMBRYONIC
MORTALITY IN THE COW

J. M. Sreenan & M. G. Diskin,

The Agricultural Institute,
Belclare, Tuam, Galway, Ireland

ABSTRACT

The extent and timing of embryonic mortality in heifers, normal cows and repeat breeder cows has been reviewed. Average fertilisation rates for normal heifers and cows were 88 and 90 percent while calving rate to a single service averaged 55 percent overall. Based on a fertilisation rate of 89 percent this is equivalent to a total level of embryonic mortality of the order of 38 percent. In normal cows, embryonic mortality would seem to occur gradually after fertilisation but with the greatest increment of loss occurring between about Days 15 to 18. Embryo losses after implantation through to full term are of the order of 5-8 percent. Reports on the extent and timing of embryonic mortality in repeat breeder cows are inconsistent. The unilateral or bilateral distribution of two embryos within the uterus had no effect on subsequent embryo or foetal mortality.

INTRODUCTION

Reproductive wastage is a major cause of production loss in both dairy and beef cow herds. Fertilisation failure and early embryonic mortality are suggested to be the main causes of this reproductive wastage. Because it is difficult to make direct measurements, many of the published estimates of fertilisation and embryo loss are based on the proportion of normal and extended return intervals following insemination. However, extended cycle lengths may occur for a variety of reasons and thus, some of these estimates should be used with caution. Embryonic mortality properly defined relates only to embryonic deaths that occur after fertilisation.

In many research programmes, greater emphasis is now being placed on attempts to increase reproductive rate in the cow by increasing the proportion of multiple (twin) births. Approaches include embryo transfer, genetic selection and steroid immunisation. The latter two approaches are aimed at increasing ovulation rate. It is pertinent, therefore, to also examine the relationship between ovulation rate, multiple ovulation distribution and embryonic mortality in the cow.

EXTENT OF EMBRYO MORTALITY

The extent of embryo mortality in single ovulating heifers and cows can be estimated from the difference between fertilisation rate and subsequent calving rate.   Direct estimates of fertilisation rate can be obtained following recovery and morphological examination of ova or embryos soon after breeding.   Fertilisation rate is measured as the number of normal cleaved ova or embryos as a proportion of all ova (fertilised and unfertilised) recovered.   Objective measurement of fertilisation rate can be made up to and including Day 8 while the ova are still within the zona pellucida.   Retention of the zona pellucida by all ova whether unfertilised, or fertilised and degenerating up to Day 8 means that there is an equal chance of recovering each ovum type and that failure to recover is of a technical nature and not due to embryonic mortality.

Published measurements of fertilisation rate are summarised in Table 1 for heifers and in Table 2 for cows. Overall estimates of 88 and 90 percent were calculated for heifers and cows respectively and with little variation between individual estimates.   These data suggest a fertilisation failure rate of about 10-12 percent.   However, because of a greater efficiency of oestrous detection under controlled conditions and the possibility of excluding animals with abnormal genitalia, the fertilisation rates recorded here probably over-estimate those achieved on farms.

While direct comparisons have not been made, there is no evidence in the literature to suggest that fertilisation rates are different following either artificial insemination with frozen-thawed semen or natural service.   This is consistent with the report of O'Farrell (1977) who recorded similar calving rates following a comparison of natural service with artificial insemination using frozen-thawed semen from the same bulls used in natural service.

It is generally accepted that calving rates to a single insemination are close to 55 percent (Diskin & Sreenan, 1980) though this may vary with age, physiological status, calving to service interval and various management factors. The difference between the average fertilisation rates of 88-90 percent calculated (Tables 1 and 2) and an average calving rate of 55 percent suggests a total pregnancy loss of the order of 33-35 percent. Based on an average fertilisation rate of 89 percent for heifers and cows, this is equivalent to a total embryonic mortality rate of the order of 38 percent.

TIMING OF EMBRYONIC MORTALITY

Ovum and embryo recovery experiments have also been employed in various attempts to measure the time at which embryonic mortality occurs and the available data are summarised in Table 3. The extent of embryo loss, both overall and at specific time intervals after insemination varies between the data sources to a greater degree than would be expected. For example, Boyd et al. (1969) record an embryo loss rate of only 8 percent by Day 25 after insemination while over the same period, Roche et al.(1981) record an embryo loss rate of 23 percent. Embryo loss estimates up to Day 42 after insemination range from 20 percent (Ayalon, 1978) to 42 percent (Diskin & Sreenan, 1980). It would seem, however, from these data (Table 3) that while embryo losses occur gradually from fertilisation onwards the greatest increment of loss would seem to occur between about Days 15 and 18. There are, however, inconsistencies between the data sources summarised, arising presumably from the variable experimental conditions employed.

After about Day 42 or from the completion of implantation onwards, most estimates in the literature suggest a foetal loss of about 5-8 percent (Boyd et al. 1969). Diskin (1984) has recorded a foetal loss of 5.5 percent based on a total of 239 cows confirmed pregnant at 50 days after insemination

REPEAT BREEDER COWS

Cows that fail to become pregnant following three or more consecutive services are classified as repeat breeder

cows and published estimates of the incidence of repeat
breeder dairy cows range from 10 to 18 percent (Ayalon, 1984).
In attempts to identify the cause of reproductive failure in
repeat breeder cows, a number of workers have examined
inter-oestrus intervals, fertilisation and embryo mortality
rates at various stages.

Reports on the timing of embryonic mortality in repeat
breeder cows are inconsistent. Hawk et al. (1955) suggests
that the major portion of embryonic loss occurs between Days
16 and 34 after insemination. On the other hand, Ayalon (1978)
suggests Day 7 as the critical day on which the major portion
of embryo deaths occur. Further Israeli data from Almeida
et al. (1984) involving embryo transfers to normal and repeat
breeder cows at six days after oestrus, records pregnancy
rates of 39 percent and 10 percent respectively. Linares
(1982) while recording a fertilisation rate of 89 percent in
repeat breeder heifers found that by Day 7, only 28 percent of
these embryos were classified as morphologically normal.
O'Farrell et al. (1983) point out that of 72 cows classified
as repeat breeders from dairy herds, more than half (55%) had
inter-oestrus intervals that were regarded as abnormal with
19 and 37 percent of return intervals either shorter than 18
days or longer than 24 days. These authors suggest that
errors in oestrous detection may be a major contributory
factor in repeat breeding. In a recent literature review,
Ayalon (1984) suggests that the major causes of reproductive
failure are fertilisation failure and embryonic deaths which,
combined, result in a level of reproductive wastage of about
twice that recorded for normal cows. On the other hand,
however, O'Farrell et al. (1983) have recorded a pregnancy
rate of 57% in cows previously classified as repeat breeders.

The repeat breeder syndrome requires to be defined and
investigated during the breeding season as there is evidence
to indicate that cows classified as repeat breeders return to
normal within a few months of the breeding season during which
they failed to become pregnant.

OVULATION RATE AND EMBRYO MORTALITY

Most oestrous cycles in the cow are accompanied by a single ovulation and little published data exists on the incidence of multiple ovulations. Following an extensive abbattoir survey, Scanlon et al. (1974) have reported an incidence of twin ovulations of 2.5 percent for beef heifers and 3.3 percent for dairy cows. Most reports concerning the incidence of twin births in cattle suggest an average of 1 - 3 percent but with considerable variation due to parity, breed and environmental conditions (Scanlon et al. 1974). That the cow is, however, capable of sustaining a significantly higher twin pregnancy rate has been shown from a series of bilateral embryo transfer experiments (Sreenan, 1977). Attempts to increase twin-pregnancy rate in the cow through an increase in the ovulation rate may, however, depend on the relationship between ovulation rate and ovulation distribution and embryonic survival rate. The level of embryonic mortality as a function of ovulation rate is higher in cows than in sheep and may be considerably higher in unilateral twin ovulating than in bilateral or single ovulating cows (Hanrahan, 1982). Using data on pregnant cows, Hanrahan, (1982) estimated that the probability of embryo loss increased by 0.22 for unilateral twin ovulators compared with bilateral or single ovulators. A previously published embryo transfer study (Rowson et al. 1972) suggested a high rate of embryo mortality where two embryos were transferred to one uterine horn, however, the number of animals involved in this study was small.

The effect of embryo distribution within the uterus on embryonic mortality is being studied and preliminary data have been published (Sreenan, 1984) and is reproduced here (Table 4). Two embryos at the morula or blastocyst stage were transferred by flank, either unilaterally or bilaterally, to synchronous recipient cows. Pregnancy, embryo survival and twinning rates were similar following unilateral and bilateral twin embryo transfers.

Data is now also available on a further group of 26

recipients that were diagnosed pregnant at 60 days after transfer and held to calve (Table 5). Twin-pregnancy and foetal survival rates to full term were similar for the unilateral and bilateral groups. These data would indicate that at least up to 65 days of gestation mortality is not increased when two embryos are confined to the one uterine horn. Further data need to be collected over the period from Day 65 to full term though the indications (Table 5) are, that embryo distribution within the uterus has little detrimental effect on foetal survival.

REFERENCES

Almeida, A.P., Ayalon, N., Faingold, D., Marcus, S. & Lewis, I. (1984). The relationship between uterine environment and early embryonic mortality (EEM) in normal (NB) and repeat breeder (RB) Friesian cows. Proc. 10th Int. Congr. Reprod. A.I. Vol. 1., 438.

Ayalon, N. (1972). Fertility losses in normal cows and repeat breeders. Proc. 7th Int. Congr. Reprod. A.I. Munich. Vol. 1.,741-744.

Ayalon, N. (1978). A review of embryonic mortality in cattle. J. Reprod. Fert. 54.,483-493.

Ayalon, N. (1984) The repeat breeder problem. Proc. 10th Int. Congr. Anim. Reprod. A.I. IV, 10pp.

Beardon, H. J., Hansel, W. & Bratton, R.W. (1956). Fertilization and embryonic mortality rates of bulls with histories of either low or high fertility. J. Dairy Sci. 39, 312-318.

Boyd, J., Bacisch, P., Young, A. & McCracken, J.A. (1969). Fertilization and embryonic losses in dairy cattle. Br. Vet. J. 125, 87-97.

Diskin, M. G. (1984). unpublished data

Diskin, M. G. & Sreenan, J. M. (1980). Fertilization and embryonic mortality rates in beef heifers after artificial insemination. J. Reprod. Fert. 59, 463-468.

Hanrahan, J. P. (1983). The inter-ovarian distribution of twin ovulations and embryo survival in the bovine. Theriogenology 1, 4-7.

Hawk, H.W., Wiltbank, J.N., Kidder, H.E. & Casida, L.E. (1955). Embryonic mortality between 16 and 34 days post breeding in cows of low fertility. J. Dairy Sci. 38, 673-676.

Henricks, D.M., Lamond, D.R., Hill, J.R. & Dickey, J.F. (1971). Plasma progesterone concentrations before mating and in early pregnancy in the beef heifer. J. Anim. Sci. 33, 450-454.

Linares, T. (1982). Embryonic development in repeat breeder and virgin heifers seven days after insemination. Anim. Reprod. Sci. 4, 189-198.

Maurer, R.R. & Chenault, J.R. (1983). Fertilization failure and embryonic mortality in parous and non-parous beef cattle. J. Anim. Sci. 56, 1186-1189.

O'Farrell, K.J. (1977). A comparison of natural service and artificial insemination in seasonally calving dairy herds. Ir. J. Agric. Res. 16, 251-257.

O'Farrell, K.J., Langley, O.H., Hartigan, P.J. & Sreenan, J.M. (1980). Fertilization and embryonic survival rates in repeat breeder dairy cows. Proc. 11th Int. Congr. of Diseases of Cattle. Tel Aviv, Oct. 1980.

O'Farrell, K.J., Langley, O.H., Hartigan, P.J. & Sreenan, J. M. (1983). Fertilization and embryonic survival rates in cows culled as repeat breeders. Vet. Rec. 112, 95-97.

Roche, J.F., Boland, M.P. & McGeady, T.A. (1981). Reproductive wastage following artificial insemination of heifers. Vet. Rec. 109, 401-403.

Rowson, L.E.A., Lawson, R.A.S. & Moor, R.M. (1971). Production of twins in cattle by egg transfer. J. Reprod. Fert. 25, 261-268.

Scanlon, P.F., Gordon, I. & Sreenan, J.M. (1974). Multiple ovulations, multiple pregnancies and multiple births in Irish cattle. J. Agric. Fish. (Dublin) 70, 1-8.

Shelton, J.N., Heath, T.D., Old, K.C. & Turnbull, G.E. (1979). Non-surgical recovery of eggs from single ovulating bovines. Theriogenology, 11, 149-152.

Smith, M.F., Nix., K.J., Kraemer, D.C., Amoss, M.S., Herron, M.A. & Wiltbank, J.N. (1982). Fertilization rate and early embryonic loss in Brahman crossbred heifers. J. Anim. Sci. 54, 1005-1011.

Spitzer, J.C., Niswender, G.D., Seidel, G.E.Jr. & Wiltbank, J.N. (1978). Fertilization and blood levels of progesterone and LH in beef heifers on a restricted energy diet. J. Anim. Sci. 46, 1071-1077.

Sreenan, J. M. (1977). Embryo transfer for the induction of twinning in cattle. In "Embryo transfer in Farm Animals" (Ed. K.J. Betteridge) (CAN. Dept. Agric.) 62-66.

Sreenan, J. M. (1984). Steroid immunisation in cows: Potential for increasing ovulation and twinning rates. Proc. 10th Int. Congr. Reprod. A.I. Urbana-Champaign. Vol. IV, 6pp.

Wishart, D.F. & Young, I.M. (1974). Artificial insemination of cattle at a pre-determined time following treatment with a potent progestagen SC-21009 . Vet. Rec. 95, 503-508.

TABLE 1   Fertilisation rate estimates following natural or artificial insemination in heifers

Reference	Days from Insemination	Ovum + embryo recovery rate (%)	Fertilisation rate (%)
Beardon et al. (1956)	3	–	$48/_{55}$ (87)
Henricks et al. (1971)	3	$18/_{20}$ (90)	$16/_{18}$ (89)
Wishart & Young (1974)	4	$43/_{50}$ (86)	$41/_{43}$ (95)
Spitzer et al. (1978)	2 – 4	$22/_{30}$ (73)	$18/_{22}$ (82)
Diskin & Sreenan (1980)	4	$30/_{35}$ (86)	$27/_{30}$ (90)
	8	$41/_{44}$ (93)	$37/_{41}$ (91)
Linares (1981)	7 – 8	$32/_{44}$ (73)	$31/_{32}$ (97)
Roche et al. (1981)	3	$52/_{60}$ (87)	$42/_{52}$ (81)
	8	$55/_{71}$ (77)	$46/_{55}$ (84)
Smith et al. (1982)	3	$45/_{57}$ (79)	$41/_{45}$ (91)
Maurer & Chenault (1983)	2 – 5	$34/_{39}$ (87)	$31/_{34}$ (91)
	6 – 8	$32/_{33}$ (97)	$24/_{32}$ (75)
All	2 – 8	$404/_{483}$ (84)	$402/_{459}$ (88)

Fertilisation rate 88%

Fertilisation failure   =   100 - 88   =   12%

TABLE 2   Fertilisation rate estimates following natural or artificial insemination in cows

Reference	Days from Insemination	Ovum + embryo recovery rate (%)	Fertilisation rate (%)
Boyd et al. (1969)	2 - 5	$33/_{46}$ (72)	$28/_{33}$ (85)
Ayalon (1972)	3	-	$10/_{12}$ (83)
	4 - 5	-	$22/_{25}$ (88)
Shelton et al. (1979)	7 - 8	$30/_{39}$ (77)	$26/_{30}$ (87)
Maurer & Chenault (1983)	2 - 5	$19/_{19}$ (100)	$19/_{19}$ (100)
	6 - 8	$17/_{17}$ (100)	$17/_{17}$ (100)
All	2 - 8	$99/_{121}$ (82)	$122/_{136}$ (90)

Fertilisation rate 90%

Fertilisation failure = 100 - 90 = 10%

TABLE 3   Fertilisation, pregnancy and embryonic survival estimates at various intervals from insemination in heifers and cows

Reference	Fertilisation rate	Pregnancy and embryo survival rates at intervals (days) after insemination					
		2 - 5	8 - 10	11 - 13	14 - 16	17 - 19	25 - 42
Beardon et al. (1965)	48/55	47/55					40/55
Boyd et al. (1969)	28/33	28/33			12/20		31/40
Henricks (1971)	16/18	16/18	2/2	13/17			12/20
Ayalon (1978)	32/37	32/37	13/18	16/18	16/20	12/21	9/13
Diskin & Sreenan (1980)	64/71	27/30	34/41	24/35	23/39		23/44
Roche et al. (1981)	88/107	42/52	46/55		49/65	47/78	18/29
Smith et al. (1982)	41/45	41/45			38/45		39/54
Maurer & Chenault (1983)	55/66	26/34			21/30		
Maurer & Chenault (1983)	36/36	15/17			14/19		
All % Pregnant	408/468 87%	274/321 85%	95/116 82%	52/70 74%	173/238 73%	59/99 60%	172/255 67%
% Embryo survival based on the fertilisation rate	100%	98%	94%	85%	84%	69%	78%

TABLE 4   Twin embryo transfers   - survival rates at Day 45 - 65

	Two embryos	
	Uni-lateral	Bi-lateral
Pregnancy rate (%)	$31/_{44}$ (73)	$20/_{26}$ (77)
Embryo survival (%)	$47/_{62}$ (76)	$32/_{40}$ (80)
Twinning rate (%)	$16/_{31}$ (52)	$12/_{20}$ (60)
"Litter size"	1.50	1.60

TABLE 5   Twin embryo transfers   - survival rates at full term

	Two embryos		Field data
	Uni-lateral	Bi-lateral	Bi-lateral
No. held to calve	18	8	206
No. calved	17	8	
No. aborted	1*	0	
No. with twins	$7/_{17}$ (41%)	$4/_{8}$ (50%)	$97/_{206}$ (47%)
Embryo survival	$24/_{34}$ (71%)	$12/_{16}$ (75%)	$303/_{412}$ (74%)
Litter size	1.41	1.55	1.47

* Aborted twins at Day 210

TIMING AND EXTENT OF EMBRYONIC MORTALITY IN PIGS, SHEEP AND GOATS: GENETIC VARIABILITY.

G. Bolet

Institut National de la Recherche Agronomique
Station de Genetique quantitative et applique
C.N.R.Z. -78350 Jouy-en-Josas, France

ABSTRACT

Embryonic mortality rate in pigs, sheep and goats ranges around 30%. It occurs mainly during the first month of pregnancy and mostly before implantation. In pigs, a second less important stage observed about Day 50 of pregnancy is resulting from overcrowding of the uterine horns. In these three species, chromosomal abnormalities such as centric fusions or reciprocal translocations cause the formation of embryos with unbalanced karyotypes which do not survive until parturition. This may considerably increase the basal loss. Several examples of major genes involved in embryonic mortality are reviewed though few cases are conclusive. The between breed and crossbred variability in embryo loss is reviewed. In pigs, large between breed variability independant of the ovulation rate is observed as well as significant individual and maternal heterotic effects. In sheep, this variability is less marked and almost no heterosis is observed. The within breed variability does not include an additive genetic component in the pig but is mainly related to the variability of the ovulation rate due both to genetic correlations and to maternal uterine effects, though the mechanism is not well understood. The success of selection experiments for prolificacy in sheep seems to be due to a lower correlation between ovulation rate and embryonic mortality than exists in the pig. The existence of pig and sheep breeds with low rates of embryonic mortality and the selection of highly prolific lines within these species provide ideal material for studies on embryonic mortality.

INTRODUCTION

Individual or herd fertility depends on a series of events ranging from gamete maturation to the birth of offspring, viz., ovulation, fertilization, egg cleavage, blastocyst implantation and embryonic and foetal development. Embryonic mortality is mainly a component of prolificacy and the definition of embryonic mortality varies according to author. Some definitions limit it to death of the embryo prior to implantation, others include foetal mortality. In this paper embryo mortality is defined as death occurring during the first

month of pregnancy.   An attempt will also be made to
determine whether the routes of improvement, especially the
genetic ones, are the same in the three species.

ESTIMATION OF THE EXTENT AND TIMING OF EMBRYONIC MORTALITY
     The embryonic mortality rate in pigs, sheep and goats
has already been widely studied and the results reported.
According to different reviews (Hanly, 1961; Lynsget, 1968;
Edey, 1969; Hughes and Varley, 1980; Flint et al., 1982;
Riera, 1982), the rate of embryonic mortality seems to be
about 30% and similar in all three species.   Mortality occurs
mostly before the fifth week of pregnancy, in pigs and sheep
(Table 1).   The values obtained are not as variable in pigs

TABLE 1:   Average estimations of embryonic mortality in pigs,
           sheep and goats

Species	Stage of pregnancy (days)	Embryonic mortality rate (%)	
Pigs	9-18	gilts :	13-18
		sows :	21-52
	17-28	gilts :	25-43
		sows :	17-39
	26-40	gilts :	38-38
	55-70	gilts :	23-50
	Farrowing	sows :	27-44
Sheep	18		11-43
	28-30		6-40
	Lambing		33-48
Goat			6-42

as in sheep and goats and this may be due to the methods of
estimation as well as to the species.   The first possible
error is an overestimation because fertilization failure may
be mistaken for embryonic mortality.   However, this error
is not important as the fertilization rate is very high in
these species according to most authors (Perry and Rowlands,
1962; Quinlivan et al., 1966; Restall et al., 1976;
Armstrong et al., 1983) and may be considered as an all or

none character, irrespective of the ovulation rate. The
second possible error is an underestimation because early
embryonic mortality may be mistaken for fertilization failure
when it occurs before the time of normal corpus luteum
regression and, therefore, without changing cycle length.
This problem is not similar in the three species. In the pig,
too few embryos (less than four according to Polge et al.,
1966) do not allow the establishment of pregnancy and leads
to a return to oestrus. This is, however, a rare phenomenon
due to either extremely low ovulation or fertilization rates.
Conversely, in sheep and goats, early mortality of all embryos
is more frequent as in these species most breeds are poorly
polytocous and this may lead to a more frequent underestimation
of the embryo mortality rate. However, in several experiments
in sheep, the slaughter of animals two days after mating
allowed accurate estimation of fertilization rate and this was
then taken into account. Edey (1969) observed that values
obtained for embryo mortality by most authors ranged from 20
to 33%, irrespective of how they were estimated. Restall
and Griffiths (1976) considered that the inclusion of
fertilization failure rate as part of embryonic mortality
had a negligible effect on the estimates of embryonic
mortality.

The most precise descriptions were made by authors
performing estimations of mortality rate at various intervals
after fertilization (Table 2). These studies confirmed that in
pigs and sheep, mortality occurs mainly during the first month
of pregnancy. In the pig, mortality was mostly observed
before the 18th day, but the variability of the slaughter times
chosen did not allow a precise timing. Nevertheless, it
seemed that mortality during the first week did not exceed
10% (Perry and Rowlands, 1962); the period prior to
implantation seemed to be the most critical stage because of
the changes in blastocyst development and their migration
before commencement of attachment (Anderson, 1974; Scofield,
1972). From the 25th day onwards only 5 to 10% of residual
mortality was observed. In sheep, Quinlivan et al. (1966)

TABLE 2.   Estimations of timing of embryonic mortality by
            successive slaughters

Species-Authors	Stage	Number of animals	Stage of pregnancy (days)	Embryonic mortality rate (%)
Pigs				
Baker et al. (1958)	gilts	24 25	25 70	25.4 42.7
Day et al. (1959)	gilts	22 10	25 40	33.3 37.9
Perry and Rowlands(1962)	gilts and sows	21 19 13 5	6 6-9 13-18 26-40	9.2 28.0 28.4 34.8
Marlowe and Smith(1971)	gilts and sows	20 20	10 25	14.7 29.8
Scofield et al. (1974)	gilts	15 15	9 13	21.4 52.4
Webel and Dziuk(1974)	gilts	34 27 39	25-30 31-40 41-112	24 17 26
Dufour and Fahmy(1975)	sows	51 42 45	23 42 63	18.3 25.8 22.2
Lutter(1981)	gilts	49 18 10	38 76 107	34.4 33.0 26.4
Sheep	parous ewes	98 86 60	18 30 140	27.0 18.1 20.0
Quinlivan et al. (1966)	non parous ewes	85 72 53	18 30 140	20.9 15.9 22.0

observed embryos dead at different stages and showed that the
period of maximal loss occurred immediately preceding the 18th
day; this accounted for 50% of the overall mortality.   Casida
et al. (1966) and White et al. (1971) demonstrated a
relationship between embryonic mortality and the magnitude of
transuterine migrations.   Moor and Rowson (1960) showed from
an egg transfer experiment that mortality was negligible (1-2%)
before the 18th day and that mortality rate was 32 to 55%
before attachment and 10 to 13% thereafter.

Many factors (breed, age, feeding, season, temperature)
may contribute to this mortality rate of 30% often called
"basal loss". The question of it's biological significance
and of ways to reduce it are of considerable importance.
This problem is even more difficult because three components
are involved i.e. the male by the "quality" of semen, the
female by the "quality" of ova and uterine environment, and
the embryo itself; in addition the pregnancy stage at which
mortality occurs is also important.   In 1964, Bishop put
forward the hypothesis that " a considerable part of embryonic
death is unavoidable and should be regarded as a normal way
of eliminating unfit genotypes in each generation .... at a
low biological cost".   However, according to this hypothesis
embryonic mortality arises mainly from unfavourable
identifiable genetic factors.   In domestic species, however the
economic costs of this mortality are rather high and, therefore
the validity of this fatalistic hypothesis should be examined
in more detail.

THE IMPORTANCE OF IDENTIFIABLE GENETIC CAUSES OF EMBRYONIC
MORTALITY

Two groups of identifiable genetic factors may affect
reproduction, viz., chromosomal abnormalities and genes with
a major effect or playing a role as markers.

Chromosomal abnormalities

Different types of chromosomal abnormalities have been
found in pigs, sheep and goats.   The main ones are aneuploid

such as Robertsonian translocations (centric fusions) and
reciprocal translocations (Bruere, 1974; Fechheimer, 1981).
These abnormalities may act at different stages of
reproduction and especially on the viability of the gametes
or on the viability of the embryos; we are only going to
study the latter case.

Aneuploidies
     Experimental heteroploids may be induced in the pig by
delaying the time of mating from the onset of oestrus.
Thibault (1959) observed 11% polyspermic eggs; Bomsel-
Helmreich (1961) found that 4.1% of embryos were heteroploid
on Day 17 of pregnancy while no chromosomal abnormalities
were found on Day 26 of pregnancy (Bomsel-Heomreich, 1965
cited by Bruere, 1974). Likewise, McFeely (1967) showed that
10% of the blastocysts collected in seven sows ten days after
mating were abnormal and were mainly polyploid. Smith and
Marlowe (1971) observed after 25 days of gestation in nine
sows that only one of 68 viable embryos was a carrier of a
chromosomal abnormality (monosomy). In sheep, Long and
Williams (1980) found 13 abnormal embryos out of 89 (14.6%)
two days post coitum of which four were trisomic. Conversely,
Long (1977) did not observe any chromosomally abnormal embryos
among preimplantation embryos.
     The centric fusions (Robertsonian translocation) represent
a particular case. These abnormalities have been observed
in cattle, in wild pigs and in some sheep breeds at a high
frequency (Bruere, 1974). In the pig, McFee and Banner
(1969) did not find any reduction in fertility or any unbalance
in the proportion of normal (38 chromosomes) and abnormal
(36 or 37 chromosomes) pigs. According to Bruere and Ellis
(1974) "there is little or no evidence to suggest that centric
fusions in a variety of combinations affect the total
reproductive fitness of domestic sheep". However, among the
three types of translocations studied by these authors some
cause a significant level of embryo loss. A centric fusion
has also been found in the goat (Soller et al., 1966; Hulot,

1969). No animal homozygous for that fusion or carrier of an unbalanced karyotype was found among the offspring of the breeding animals carrying that abnormality (Padeh et al., 1971; Popescu, 1972; Ricordeau, 1972).

Reciprocal translocations

An unfavourable effect of a reciprocal translocation on prolificacy has been recorded in sheep (Glan-Lugt and Wabmuth, 1980), resulting in an increase in embryonic mortality. Matings between carriers of the translocation resulted in a reduction (10-40%) in the number of twins and triplets. According to the review of Popescu et al. (1984), 11 types of reciprocal translocations have been demonstrated in the pig causing a 26 to 100% reduction of litter size. Akesson and Henricson (1978) studied the progeny of a boar heterozygous for a reciprocal translocation and found that the fertility was reduced by 30% as compared to a normal boar. Five different types of unbalanced karyotypes were found in the embryos and foetuses produced by this boar but none were found in the piglets born. Forty one percent were carriers of the translocation. The same observation was made by Popescu and Boscher (1982), on Day 10 of gestation the proportion of embryos with normal or abnormal karyotypes was in keeping with the expected values. It may, therefore, be concluded that the reduction in litter size is due to embryonic death of individuals with unbalanced karyotypes occurring at different stages of gestation (King et al., 1981).

Thus, chromosomal abnormalities, especially centric fusions and reciprocal translocations have an unfavourable effect on prolificacy arising from their effects on embryonic mortality and not from their effects on the gametes (as shown in mice by Ford and Evans, 1973). The increased mortality due to the elimination of unbalanced karyotypes has variable and sometimes considerable economic consequences according to the frequency of these abnormalities (Popescu et Tizier, 1984). This justifies the setting up of an eradication programme, especially a systematic examination of the karyotypes of the males used in artificial insemination.

Major genes and genetic markers

In the pig, the main markers studied are blood group H, transferin locus, the major histocompatibility complex (SLA) and the halothane sensitivity gene (reviewed by Olliver and Sellier, 1982). The published results are not consistent, however, and it is not always possible even when an effect is significant, to define at which level of the reproductive process this effect operates. This is particularly true when authors describe the effect of an allele present in the female viz., the unfavourable effects of allele Ha (Jensen et al., 1968), or some SLA haplotypes (Schworer et al., 1984), or of the halothane sensitivity gene (Willeke et al., 1984). Conversely, other authors described the effect of specific sire x dam combinations on litter size and on the proportion of specific embryo genotypes. This is most likely the effect of an allele on embryonic mortality. Thus, Rasmusen and Hagen (1973) observed in the Yorkshire breed a decrease in litter size and in the proportion of Ha negagive piglets. Our studies on SLA (Renard et al., 1985) have demonstrated a deficit of piglets homozygous for a haplotype (8.9% instead of 25%) together with a large reduction in litter size (-2.8) at birth (P<0.01). Although no definitive conclusion can be drawn from this data, this effect may be due to the presence of a lethal homozygous gene combined with SLA. Webb and Jordon (1979) showed that litter size was significantly reduced (-1.6) when all piglets were halothane sensitive, and thus homozygous for that gene, but that the sensitivity of the dam to halothane did not have a significant effect (-0.4). This is in keeping with the results of Schworer and Morel (1984) showing that the size of litters which were 100% heterozygous for that gene, was higher than that of litters which were 50% heterozygous (+0.5) or 100% homozygous (+0.9). This advantage of heterozygous litters is also in agreement with the results of Rasmusen and Hagen (1983) concerning blood group H and results from this laboratory (Renard et al. 1985) concerning SLA. This may due to a better survival rate of embryos heterozygous for some genes. This phenomenon may also be related to the immunological relationship during gestation between the dam and the embryo due to their genotypes (see Beer and Sio, 1982).

In sheep, Ricordeau (1982) reported the effect of several types of genetic markers on fertility. However, it is only possible to impute this action as an effect on embryonic mortality when the types of mating and the genotype of the female are studied. The lethality of several coloration genes is also reported by Schinckel (1955) and Ricordeau, (1982). The gene responsible for the presence of wattles seems to affect prolificacy, but Casu et al. (1970) did not observe any unbalances in the frequency of the different genotypes among the progeny. Haemoglobin type also has an effect with Hba having an unfavourable effect. According to Arora et al. (1971), it is connected with a higher embryonic mortality especially in AA homozygotes and Walker et al.(1981) did not exclude such hypothesis. The gene responsible for the prolificacy of Booroola Merino sheep seems to depend mainly on it's additive effect on ovulation rate (Bindon, 1984). However, the data also suggest the presence of a dominance effect on embryonic survival (Davis et al., 1982) (Table 3).

In the goat, the main case known is a dominant autosomal gene P leading to polledness. In matings where PP homozygotes are expected there is a significant female deficit even when considering hemaphrodites as females (Soller and Angel, 1964). The available data, however do not allow the authors to choose between the hypothesis of sex-reversal or lethality. There are also cases of over-prolificacy depending on the presence of allele P, heterozygous in the female (Soller and Kempenich, 1964) and homozygous in the male (Ricordeau, 1969), but the mechanism has not yet been determined. Thus, the possibility that gene P exerts an action, either favourable or unfavourable on embryonic mortality cannot be excluded. However, it's main action is a recessive effect leading to full sterility in the females and partial sterility in the males)Lauvergne, 1969), thus justifying the setting up of an eradication programme (Ricordeau and Sanchez, 1981).

TABLE 3.   Effects of Boorola gene (F) on ovulation rate and embryonic mortaltiy (Davis et al., 1982)

Genotype	Sire X Dam	n	Ovulation rate	Embryonic mort. (%)
F/F	F/+ X F/+	20	4,59	41,0
F/+	F/+ X F/+	61	2,78	19,4
	F/+ X +/+	27	2,78	18,7
	Overall	88	2,78	19,2
+/+	F/+ X F/+	38	1,50	10,7
	F/+ X +/+	28	1,48	6,8
	Overall	66	1,49	9,0
2F+	F/+ X F/+		-9%	-42%

According to this review, the importance of identifiable genetic factors in pigs, sheep and goats does not seem as marked as in man.   In some cases, they may lead to a considerable increase in embryonic mortality, but they do not fundamentally account for the basal loss.   Nevertheless, other genetic explanations may be involved and identified through the study of between and within-breed variability.

BETWEEN-BREED VARIABILITY IN EMBRYONIC MORTALTIY

Comparison between breeds or crosses involves the genotype of the dam, the sire and the embryo.   Four types of parameters may be used to estimate their value, viz., the mean effect of pure breeds, the individual heterotic effect on the crossbred embryos, the maternal or paternal heterotic effect due to the crossbred female or male.   As the comparison is often made between breeds with different ovulation rates, it is important not to mistake the effect of ovulation rate on embryonic mortality for the breed's own uterine efficiency

effect. The importance of that problem is not the same in
the three species.

Pigs

The ovulation rate varies from 12 to 20 (corpora lutea)
and the embryonic mortality rate from 15 to 40% (Table 4) in
the main breeds. Comparison of pure breeds shows that there
is no close relationship between these two parameters.

TABLE 4.   Some comparisons of ovulation rate and embryonic
mortality in pig breeds. (The number with the same
letter are not significantly different).

Authors	Breeds	No. of sows	Ovulation rate		No. of embryos		Embryonic mortality (%)	
					(30 days)			
Young et al.	Duroc	25	14,1	a	11,1	a	19	a
(1976)	Hampshire	28	12,4	b	8,4	a	30	b
	Yorkshire	19	13,9	a	11,2	a	16	a
					(80—110 days)			
Hagen and Kephart (1980)	Feral swine	17	8,7	a	6,2	a	29	a
	Yorkshire	12	15,1	b	9,8	b	35	a
					(30 days)			
Legault and Gruand (1981)	Large White	415	15,4		9,4		39	
	Landrace	110	13,7		9,6		30	
Bolet and Martinat-Botte (1984)	Large White	13—20	18,9	a	11,9	a	31	a
	Meishan	16—17	17,5	a	14,8	b	14	b

Thus the ovulation rate of the feral swine is 42% lower than
that of the Yorkshire pig, but the embryonic mortality is
almost the same according to Hagen and Kephart (1980).
However, the mortality rate of the former is only 13% according
to Aumaitre et al., (1984). Likewise, the ovulation rate of
the very prolific Meishan sow is almost the same as that of the
Large White sow, but the embryonic mortality rate is very low
(Bolet and Martinat-Botte, 1984). Conversely, in another

prolific chinese breed, the Jiaxing, both the ovulation rate
and embryonic mortality seem to be high (observations on
crossbreds, Rombauts et al. 1982). Among the main European
breeds, the embryonic mortality rate of the Large White sow
seems to be much higher than that of the Landrace sow, but
the ovulation rate seems also to be much higher (Legault and
Gruand, 1981). As far as American breeds are concerned, the
Hampshire sow seems to have a low ovulation rate and a higher
embryonic loss rate than the Duroc and Yorkshire sows (Young
et al., 1976).

Sellier, (1976, 1982) and Johnson (1981) have published
a review of the individual and maternal heterotic effects on
reproductive traits. The heterotic effect (maternal) on the
ovulation rate ranged from low to non-existent (0.1-2%). The
effects of heterosis on the number of live embryos or on litter
size at birth was larger. The individual heterosis ranged
from 3 to 5% and the maternal heterosis from 7 to 10%, but
this estimation included part of the individual heterosis.
There does not seem to be any signficant paternal heterotic
effect on prolificacy (Sellier, 1982). However, these values
vary according to the paternal and maternal breed combinations
and according to their purebred means. Thus, Legault and
Caritez (1983) observed a heterotic effect on litter size of
12 to 15% in Chinese X European crossbred sows. This was
higher than that usually observed in Large White X Landrace
crossbred sows where the effect was low (Sellier, 1982).
According to several investigations, this improvement in
litter size from crossbreeding was due to a reduction in the
embryonic mortality rate ranging from 14% (Legault and
Gruand, 1981) to 28% (Johnson et al. 1978). The estimates
showed that this favourable effect was due to the improvement
of embryo viability and/or of the maternal qualities of the
crossbred sow. Only embryo transfer experiments would
partition embryo genotype effect from that of the dam on the
one hand and the ovulation rate effects from uterine capacity
on the other.

Sheep

The variability of litter size in sheep both between and
within breeds is definitely larger than in pigs.   According
to Bradford (1972), there may be a two-fold difference in
litter size between breeds kept in the same environment.
This estimate did not include more prolific breeds, Romanov,
Finnish Landrace, Booroola Merino line or the highly prolific
Chinese breed Hu-Yang (Epstein, 1969).   This reflects the
variability in ovulation rate, ranging from 1 to 2.5 corpora
lutea for most breeds (Bradford, 1972) to 2.5 - 4 for Romanov
(Ricordeau et al. 1982) and Finnish Landrace (Lawson and
Rowson, 1972;   Hanrahan and Quirke, 1977).

Embryonic mortality rate varies between and within breeds
according to ovulation rate.   For that reason, comparisons
between breeds or crosses should be made at the same ovulation
rate using a selected sample.   Only few comparisons have
been made in that way and these generally do not show clear
differences (Bradford, 1972).   Ricordeau et al.(1972. 1982)
showed that the embryonic mortality rate is lower (-20%) in
Romanov than in Berrichon due Cher and Finnish ewes.   Meyer
and Clarke, (1982) did not find any difference in uterine
capacity between Romney and Border Leicester breeds.   A great
number of embryo transfer experiments have been carried out.
Using that technique it is possible to compare breeds by
standardizing the number of implanted embryos and by
partitioning the genyotye of the dam from that of the embryo
(Table 5).   According to these experiments there is not an
effect of embryo genotype on the embryonic mortality rate.
Most differences observed between recipient dams were also
non-significant.   Only Lawson and Rowson (1972) observed a
significantly lower embryo mortality rate in Finnish ewes; a
similar although non significant trend was observed by Hanrahan
and Quirke (1977).   Bradford et al.(1974) did not notice any
superiority of the Finnish breed as compared to all the other
breeds of recipient ewes.   The embryo transfer experiments
mentioned do not include the Romanov breed and, therefore, do
not allow verification of the favourable effect of the Romanov
breed on embryo survival,indicated by Ricordeau et al. (1976,

TABLE 5.   Egg transfer experiments in sheep

Authors	Breed of recipient	Mean natural ovulation rate	Breed of eggs	No. of recipient ewes lambing	No. of eggs	Embryonic mortality (%) (ewes lambing)	Embryonic mortality (%) (ewes transferred)
Larsen and McDonald (1971)	Romney	1.06		29	3	38	
	B. Leicester X Romney	1.40		25	3	29	
Moore (1968)	Merino	1.27	Merino	11	3	42} 43	
			B. Leicester	13	3	44)	
	B. Leicester	2.25	Merino	11	3	30} 26	
			B. Leicester	12	3	22}	
Bradford	Welsh Mountain	1.45	Finnish	14	4	39) 36	39
			Southdown	5	4	25)	25
	B. Leicester	1.93	Finnish	6	4	21) 27	41
			Merino	5	4	35)	35
	Finnish	3.33	Soay	5	4	35)	25
			Finnish	4	4	25) 28	25
	Finnish		Oxford	4	5	31)	31
			& Soay	4	5	30	30
				8	6	35	35
Lawson & Rowson (1972)	Suffolk	2.1	Clun Forest	12	5	43	58
	Finnish Landrace	2.6	Clun Forest	14	5	33	37
	Romney Marsh	1.4	Clun Forest	13	5	46	53
	"		Romney Marsh	8	5	50	67
	"		Finnish Landrace	8	5	40	56
Hanrahan and Quirke (1976)	Finnish Landrace Galway		Finn X Galway	}	3	33	33
	Finn X Galway		"	} 112	3	37	37
			"	}	3	33	33
Hanrahan and Quirke (1977)	Finnish Landrace Galway	3.8	Galway	10	6	64	64
	Galway	1.7	Galway	25	6	72	72

1982).   However, it may be assumed that the prolificacy of
this breed can be partly explained by this favourable effect
on embryo survival in contrast to other prolific breeds such
as Finnish or D'man (Lahlou-Kassi and Marie, 1981).

According to the reviews of Turner (1969) and Bradford
(1972) there are not any significant heterotic effects on
ovulation rate or litter size.   Only a few experiments
involving egg transfer of different genotypes have included
crossbred embryos.   Larsen and McDonald (1971) did not
observe any survival difference between Romney and Border
Leicester X Romney embryos.   Hanrahan and Quirke (1976)
transferred Finnish X Galway eggs and obtained a mortality
rate of 33 to 37% for three eggs transferred per recipient.
Bradford (1972) did not observe any significant effects of
the male breed.   Accordingly, it seems that the direct
heterotic effect on embryo viability is very low and perhaps
non-existent.   Ricordeau et al. (1972) found a maternal
heterotic effect on embryo survival in Romanov X Berrichon du
Cher ewes while Meyer and Clarke (1982) reported non-significant effects in Border Leicester X Romney crossbreds.

Goats

To our knowledge, there are only a few studies available
comparing the prolificacy and probably none comparing the
embryonic mortality of goat breeds or crosses.   The goat,
like the sheep is a poorly polytocous species, exhibiting
large variability in litter size.   The litter size of most
breeds lies between 1 and 2 (Riera, 1982, Shelton, 1978) and
the ovulation rate ranges from 1-2 in most breeds but up to 4
in some (Rao and Bhattacharya, 1980).   Some breeds seem to
be more prolific, especially in China (Jun-Quian, 1982).
Chawla and Bahatnager (1982) compared the prolificacy of
Beetal, Alpine, Saanen goats and their crossbreds.   They noted
the superiority of Beetal (litter size 1.7) but did not report
any heterotic effect.   Similar observations in sheep suggest
that the main factor contributing to variation between breeds
is the ovulation rate.   Nevertheless, the characteristics of

gestation are not the same in sheep and goats, the latter being dependent on the corpus luteum. It seems that the high abortion rate in Angora goats (Wentzel, 1982) may be attributable in some instances to corpus luteum failure.

It may be concluded that the importance of embryonic mortality is different in the three species (especially sheep and pigs) although the rates are almost the same. In pigs, the variability of embryonic mortality between breeds or crosses is larger than that of the ovulation rate whereas the reverse is found in sheep. . Consequently, choice of pig breeds or crosses is very important in obtaining a significant reduction in embryonic mortality rate, whereas it is less important in sheep compared to the role of ovulation rate. However, this conclusion concerning the between breed variability has not automatically been taken into account in the criteria of selection, which depend on the within breed variability.

WITHIN-BREED VARIABILITY IN EMBRYONIC MORTALTIY

Variability in embryonic mortality and selection on prolificacy

The efficiency of selection depends on the variability of the traits selected for in the population and on their genetic expression. The two main traits used in programmes for genetic improvement of prolificacy in sheep and pigs are ovulation rate and litter size. The results obtained vary according to the species.

In the pig, two long-term selection experiments have been carried out, one in Nebraska on ovulation rate (Cunningham et al. 1979) the other in France on litter size (Olliver and Bolet, 1981). The first one led to a signficant increase in ovulation rate (about 0.4 ova per generation) but no correlated response for litter size was found. The second experiment did not lead to any significant improvement in litter size, but led to a non-significant improvement in ovulation rate (0.2 ova per generation). In these two cases, an increase in embryonic mortality may explain the failure to improve litter size.

In sheep, several experiments concerning selection for litter
size have been carried out and reviewed by Turner (1959),
Bradford (1972) and Hanrahan (1982).   The results obtained
indicate an average increase of 0.02 lambs per ewe per year,

Genetic parameters

The heritability values for ovulation rate and litter
size in pigs (review by Johansson, 1981, Bolet and Legault,
1982) and in sheep (review by Turner, 1969, Hanrahan, 1982)
are very similar.   The heritability of litter size was
estimated to be 0.10 with estimates varying from 0 to 0.25.
Similar values have been reported for the goat by Moulick
et al. (1965).   The repeatabilities recorded by the same
authors were also close to 0.15.   Conversely, heritability
and repeatability estimates for ovulation rate were higher
(0.30-0.45).   Because of the within breed variability of
litter size in these species, the low heritability values for
that trait was compatible with the annual genetic gain of 1 to
3% obtained in sheep and expected in pigs (Olliver, 1982).

The embryonic mortality parameters have not been as
thoroughly examined as have ovulation rate and litter size.
However, it may be assumed that it's additive genetic
variability is almost zero.   Young et al (1977) calculated
that the heritability of embryonic survival rate was zero in
the pig, and Legault and Gruand (1981) though not calculating
it directly, drew the same conclusions.   In sheep, Hanrahan,
(1982) found that the repeatability of survival rate was not
significantly different from zero (0.07).   However, a direct
effect of the male on litter size was reported in the pig
following both natural mating (Uzu, 1979) and artificial
insemination (Olliver and Legault, 1967, Willeke, 1981).
The differences between males reached two piglets, excluding
the boars which were carriers of translocations (Willeke and
Richter, 1981).   In sheep, Bradford (1972) considered that
"differences between rams within group may account for a
significant amount of variation in litter size, but to date
there has been little experimental work to establish whether
such individual differences are genetic".   In males with good

semen quality this direct effect may be explained by an effect
on embryo survival rate. To our knowledge, the genetic
component of this male effect and the possibility of using it
in a genetic improvement plan have not been studied. Yet
this effect is in agreement with the already mentioned theory
of Bishop (1964) and should be more thoroughly examined.

The importance of the non additive genetic component of
embryonic mortality rate in pigs is supported by the existence
of unfavourable heterotic effects due to inbreeding. In
sheep, the direct heterotic effect is low or even non-existent.
Inbreeding has a depressive effect (Doney and Smith, 1968),
but Lax and Brown (1968) showed that this effect depends
mainly on the genotype of the dam and not on that of the
embryo. Thus, the genetic components of embryonic survival
may be different. However, this is not sufficient to explain
the discrepancy between selection results in pigs and sheep.
Indeed, selection experiments in mice, with ovulation rates
resembling more that of pigs than sheep, have led to a
significant improvement in litter size, mainly by increasing
the ovulation rate, but also by reducing the embryonic
mortality rate. These experiments in mice have demonstrated
that as in pigs, the dominance of genes affecting pre-natal
survival are additive to the genes affecting ovulation rate
(Bradford, 1979).

Selection intensity - Selection for "extremes"

Olliver and Bolet (1981) partly explained the failure
of their experiment concerning prolificacy by the failure to
reach the expected selection intensity. This problem can be
overcome by applying intense selection to a large population.
Application of this theory to the whole controlled French
pig herd led Legault and Gruand (1976) to create a hyperprolific Large White line. The utilization of daughters of
hyperprolific sows and the dissemination of this line through
the males used in AI has led to a non-significant improvement
in litter size of adult sows of one piglet and in ovulation
rate of 1.6 corpora lutea in one generation (Bolet and Legault,
1982). Conversely, in the nulliparous daughters of these

males, ovulation rate was also improved by 1.8 ova, but the
number of live embryos at Day 30 of pregnancy was not increased
due to a higher rate of embryonic mortality (Legault et al.
1981). In hyperprolific sows themselves, a significant
increase in embryo mortality rate was also observed, but it
was not large enough to offset the increased ovulation rate
(Bolet and Martinat-Botte, 1984) (Table 6). Since then this
method has been used with good results for the pig, (Bichard
and Tomkins, personal communication; Tomes and Newman, 1984).
These results, therefore, are similar to those obtained in
other species where several successful experiments were
carried out starting with highly prolific animal studies
such as in sheep (Bradford, 1972, Land, 1978) and in mice
(Bakker et al. 1978).

Genetic improvement of embryo survival

As shown above, selection at least in sheep and multiparous sows may lead to an increase in ovulation rate and litter
size. However, it is generally accompanied by a higher
level of embryonic mortality. Because of the mainly non-
additive genetic components of embryonic mortality, selection
for that trait should not in fact be efficient. Such
selection was, however, found to be efficient in mice according
to Bradford (1979). This possibility has not yet been
examined in pigs, but a series of experiments was initiated
recently. One involving selection only on embryo survival
(Webb, personal communication) and the other involving
selection on an index combining both ovulation rate and
embryo survival (Johnson et al. 1984). Another and
theoretically more efficient method of genetic improvement
would be the utilisation of crossbred females originating
from lines selected on ovulation rate, with the aim of
improving the embryo survival rate. However, the results
of Johnson et al. (1984) on crossbred females from the
Nebraska line selected for ovulation rate did not support
that hypothesis. Embryonic mortality rate was 0.18 in the
selected crossbreds and 0.20 in the controls. Legault et al.

(1981) observed that the number of live embryos at 30 days of
gestation was higher than in the crossbred controls, but this
difference was not significant (Table 6).

TABLE 6.   French experimental results on the hyperprolific
            Large White line

| Variable | Hyperprolific sows (1) | Daughters of hyperprolific boar and/or sow (2) and (3) | | |
		purebred gilts	crossbred gilts	purebred sows
Ovulation rate				
. value	23,1 (18)	16,3 (112)	15,4 (71)	18,0 (27)
. C	+4,2 **	+1,8 **	+0,2 n.s.	+1,5 **
Litter size at farrowing				
. value	13,9 (25)	9,5 (42)	–	11,3 (36)
. C	+2,0 n.s.	−0,5 n.s.	–	+0,7 n.s.
Number of embryos at 30 days of pregnancy				
. value	–	9,6 (74)	11,2 (60)	14,0 (25)
. C	–	+0,1 n.s.	+0,8 n.s.	+1,5 *
Embryonic mortality (%)	at farrowing	at 30 days pregnant		
. value	42 (17)	41 (74)	27 (60)	22 (25)
. C	+11 *	+7	−4	0

According to   (1) Bolet and Martinat Botte (1984)
                (2) Legault et al (1981)
                (3) Bolet and Legault (1982) (with 1 generation added)

        C   :   deviation from control Large White Gilts or sows
                n.s.: no significantly different from 0

            *   :   P<0,10)
            ** :   P<0,05) significantly different from 0

            ( ) :   Number of data for each variable

Thus, it is not possible to draw conclusions from the
superposition of addivive (selection) and non additive
(heterotic) effects of the genes acting on litter size.
Because of the absence of significant heterotic effects in
sheep, crossbreeding does not seem to give many advantages.
A dominance effect on litter size combined with an additive
effect on ovulation rate seems to exist only in the case of
the Booroola gene (Davis et al., 1982) Table 3).

Embryonic mortality and ovulation rate

According to the results presented above, the use of
selection programmes to capitalise on the variability that
exists for embryonic mortality are unlikely to be successful.
This may either be due to the absence of any highly correlated
measurable genetic traits upon which one may act, or to the
presence of undesirable genetic correlations, especially with
ovulation rate.   Many authors have shown a within-breed or
within-line relationship between ovulation rate and embryonic
mortality.

In pigs the different selection methods used to increase
ovulation rate have also increased embryonic mortality rate at
30 days of pregnancy (Legault et al. 1981) and at farrowing
(Cunningham et al., 1979; Bolet and Martinat-Botte, 1984)
(Table 6).   Several authors have found a curvilinear
relationship between ovulation rate and embryonic mortality
rate (Swierstra and Dyck, 1976, Legault and Gruand, 1981,
Lutter et al. 1981, Blichfeldt and Almlid, 1982, King and
Williams, 1984).   (Figure 1.)   Conversely, an experimental
increase in the number of embryos by embryo transfer did not
lead to a significant rise in the embryonic mortality rate
after 30 days of pregnancy (Pope et al. 1972). Superovulation
or an experimental overcrowding in a uterine horn (Dzuik, 1968,
Fenton et al., 1970) did not have any significant effect at 30
days of pregnancy.   Thus, it seems that uterine crowding is
not a limiting factor in embryo survival during the first
month of pregnancy.   Accordingly, this relationship does not
exclusively depend on the uterine environment, but may be

connected with the genotype of the dam or the embryo, and thus
indicate an unfavourable genetic correlation.   Overcrowding
of the uterus does, however, become a limiting factor during
the subsequent stages of pregnancy.   Johnson et al. (1984)
observed that the increase in embryonic mortality in the
Nebraska line selected for ovulation rate was due to an
increase in the rate of loss from the 50th day of pregnancy
onwards.   Polge (1982) also noticed that litter size at
farrowing had not generally been increased above normal either
by transferring additional embryos or by superovulation.
The phenotypic correlation between ovulation rate and number
of embryos ranges from 0.25 to 0.30 (Legault and Gruand, 1981,
Hohnson et al., 1984) and the genetic correlation is very low
or non-existent (Cunningham et al., 1979, Young et al., 1978.
Johnson et al., (1984) set up a selection index based on
phenotypic and genetic correlations between ovulation rate
and embryo survival of -0.4 and -0.75, respectively.   However,
these mean coefficients do not allow expression of the
complexity of the relationships between ovulation and embryonic
mortality rates.   Thus, despite the absence of a correlated
response to selection on ovulation rate for nine generations
in the Nebraska line, Johnson et al. (1984) observed that
the litter size of the selected line became significantly
larger than that of the control line after five generations of
"relaxed" selection (+0.78$^{\pm}$0.22 piglet).

In sheep, the same relationship between ovulation rate
and embryonic mortality rate was found in the Romanov breed
(Ricordeau et al., 1976 and 1982) and Finnish and Galway
breeds (Hanrahan, 1980).   The same relationship was also
evident from embryo transfer experiments (Figure 2).   Hanrahan
(1982) showed that the improvement in litter size obtained
during selection experiments in sheep depended only on an
increase in the ovulation rate and that improvement in embryo
survival had not contributed to the observed response.
However, according to Hanrahan, these selection experiments
did not cause any significant rise in embryonic mortality rate
either.   Because of these facts, it is considered that

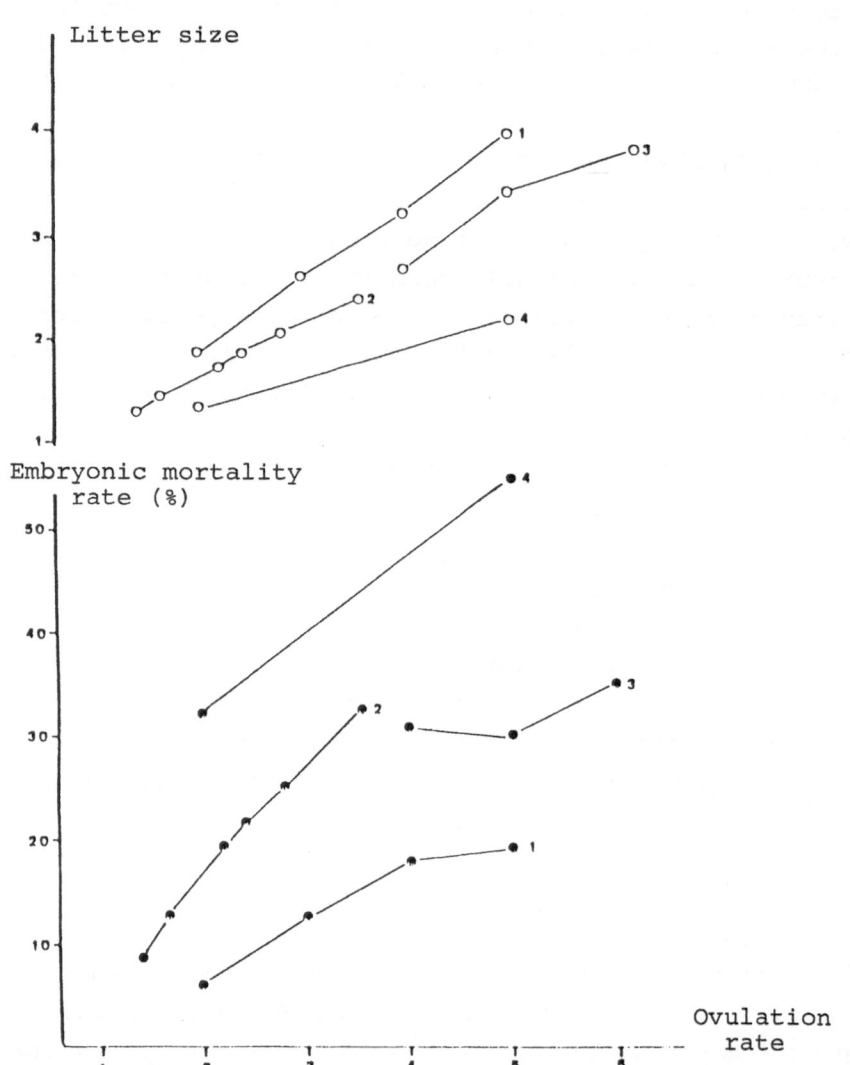

FIG. 2:   Relationship between ovulation rate and litter
           size or embryonic mortality in sheep

1 -   Ricordeau et al., 1982
2 -   Hanrahan, 1982
3 -   Bradford et al., 1974 (egg transfer)
4 -   Moore and Rowson, 1960 (egg transfer)

ovulation rate is the limiting factor to prolificacy in sheep. Indeed, as shown by Hanrahan (1982), over the range of 1.4 - 2.5 for mean ovulation rate, the change in litter size is very nearly linear at approximately 0.63 lambs per unit change in ovulation rate.   This rate is even higher (about 0.74) in the Romanov breed (Ricordeau et al., 1982) owing to better embryo survival in that breed.   However, the relationship between ovulation rate and litter size is not fully linear and the correlation between these two traits decreases with increasing ovulation rate (Hanrahan, 1980).

In goats, only a few experiments on this topic have been carried out.   It seems that an increase in embryonic mortality rate results from an increase in ovulation rate occurring from the second ovum shed (Nair and Raja, 1973, Lyngset, 1968) or the second ovum transferred (Moore and Eppleston, 1979).   However, Armstrong et al. (1983) observed that embryo survival rate was higher following the transfer of two embryos compared to one, and that the survival rate also increased with the ovulation rate of the recipient goat. Pregnancy in the goat involves a close relationshp between corpora lutea and embryos, suggesting that the relationship between ovulation rate and embryonic mortality are more complex than in sheep.   Although the genetic improvement of prolificacy is not a priority goal among the breeders, the development of embryo transfer should lead to new research programmes on reproduction in the goat (Corteel et al., 1982).

CONCLUSIONS

Some genetically identifiable factors may on occasion cause a significant increase in embryonic mortality, but they cannot explain much of the basal loss and in that respect we do not agree with the hypothesis of Bishop (1964). Neither is there much data indicating reduced mortality by genetic selection.   However, between breed variability does exist which is larger in pigs than in sheep.   Some breeds such as Meishan pigs or Romanov sheep are characterised by a low embryonic mortality.   There is also a more marked genetic

component of embryonic mortality in pigs than in sheep.
However, the nature of this component and the complexity of
the interactions with other traits such as ovulation rate
have not yet been well established. Unfortunately, some of
the hypotheses studied have not been verified by embryo
transfer studies in pigs. However, it may be assumed that
in the pig the between and within breed variability should be
used as a means of reducing the extent of embryonic mortality.
In sheep, the results of embryo transfer experiments and the
success of several selection experiments on prolificacy show
the predominant role of ovulation rate as a determining factor.
However, because breeds such as Romanov exist and also because
of the effect of the Booroola gene on embryonic mortality, the
improvement of embryo survival should be maintained as
objective research.

Improvement of prolificacy not only in pigs and sheep
but in all domestic species will in the future be based more
on a reduction in embryonic mortality.

ACKNOWLEDGEMENTS

I wish to thank Annick Bouroche and Kirsten Rerat for
translation of the manuscript, Irene Cabourdin for the
type-writing, and Denise Boyajean, P. Pospecu, G. Ricordeau,
and J. M. Corteel for their help in the literature search.

REFERENCES

Akesson, A., Henricson, B. 1972. Embryonic death in pigs caused by unbalanced karyotype. Acta Vet. Scand., 13, 151-160.

Anderson, L.L. 1974. Early embryonic development in the pig. J. Anim. Sci., 39, 986 (abstr).

Armstrong, D.T., Pfitzner, A.P., Warnes, G.M., Seamark, R.F. 1983. Superovulation treatments and embryo transfer in Angora goats. J. Reprod. Fert., 67, 403-410.

Arora, C.L., Acharya, R.M., Kakar, S.N. 1971. A note on the association of haemoglobin types with ewe and ram fertility and lamb mortality in Indian sheep. Anim. Prod., 13, 371-373.

Aumaitre, A., Quere, J.P., Peiniau, J. 1984. Influence du millieu sur la reproduction hivernale et la prolificite de la laie. Symposium international sur le Sanglier, Toulouse, 24-26 Avril, 1984, les colloques de l'INRA, 22, 69-78, INRA publ., Paris.

Baker, L.N., Chapman, A.B., Grummer, R.H., Casida, L.E. 1958. Some factors affecting litter size and fetal weight in purebred and reciprocal-cross matings of Chester White and Poland China swine. J. Anim. Sci., 17, 612-621.

Bakker, H., Wallinga, H.H., Politiek, R.D. 1978. Reproduction and body weight of mice after long-term selection for large litter size. J. Anim. Sci., 46, 1572-1580.

Beer, A.E., Sio, J.O. 1982. Placenta as an immunological barrier. Biol. Reprod., 26, 15-27.

Bindon, B.M. 1984. Reproduction biology of the Booroola Merino sheep. Aust. J. Biol. Sci., 37, 163-189.

Bishop, M.W.H. 1964. Paternal contribution to embryonic death. J. Reprod. Fert., 7, 383-396.

Blichfeldt, T., Almlid, T. 1982. The relationship between ovulation rate and embryonic mortality in gilts. Theriogenology, 18, 615-620.

Bolet, G., Legault, C. 1982. New aspects of genetic improvement of prolificacy in pigs. 2nd World congress on Gneetics applied to livestock production, Madrid, 4-8 Oct. 1982, 5, 548-567, Editorial Garsi, Madrid.

Bolet, G., Martinat-Botte Francoise, 1984. Comparison of ovulation rate and litter size of hyperprolific and control Large White sows, and Meishan sows. (to be published)

Bomsel-Helmreich Ondine, 1961. Heteroploidie experimentale chez la truie. IVth international congress on animal reproduction, the Hague, 5-9 June, 1961, 3, 578-581, N.V. Drukkerij Trio, The Hague

Bradford, G.E. 1972. Genetic control of litter size in sheep. J. Reprod. Fert., suppl. 15, 23-41.

Bradford, G.E. 1979. Genetic variation in prenatal survival and litter size. J. Anim. Sci., 49, (suppl. 2) 66-74.

Bradford, G.E., Taylor, ST.C.S., Quirke, J. F. & Hart, R. 1974. An egg transfer study of litter size, birth weight and lamb survival. Anim. Prod., 18, 249-263.

Bruere, A.N. 1974. The discovery and biological consequences of some important chromosome anomalies in populations of domestic animals. 1st World Congress on genetics applied to livestock production, Madrid, 7-11 Oct. 1974, 1, 151-175. Editorial Garsi, Madrid.

Bruere, A.N., Ellis, P.M. 1979. Cytogenetics and reproduction of sheep with multiple centric fusions (Robertsonian translocation). J. Reprod. Fert., 57, 363-375.

Casida, L.E., Woody, C.O., Pope, A.L. 1966. Inequality in function of the right and left ovaries and uterine horns of the ewe. J. Anim. Sci., 25, 1169-1171.

Chawla, D.S., Bhatnagar, D.S. 1982. Reproductive performance of dairy goats. 3rd Intern. conference on goat production and disease, Tucson, 10-15 Jan. 1982, 304, Dairy Goat Journal Publishing Co. Scottsdale, U.S.A.

Corteel, J.M., Gonzalez, C., Nunes, J.F. 1982. Research and development in the control of reproduction. 3rd Intern. conference on goat production and disease, Tucson, 10-15 Jan, 1982., 584-601, Dairy Goat Journal Publishing Co. Scottsdale, U.S.A.

Cunningham, P.J., England, M.E., Young, L.D., Zimmerman, D.W. 1979. Selection for ovulation rate in swine:correlated response in litter size and weight. J. Anim. Sci, 48, 509-516.

Davis, G.H., Montgomery, G.W., Allison, A.J., Kelly, R.W., Bray, A.R. 1982. Segregation of a major gene influencing fecundity in progeny of Booroola sheep. New Zealand J. Agric. Res., 25, 525-529.

Day, B.N., Anderson, L.L., Emmerson, M.A., Hazel, L.N., Melampy, R.M.1959 Effect of estrogen and progesterone on early embryonic mortality in ovariectomized gilts. J. Anim. Sci., 18, 607-613.

Doney, J.M., Smith, W.F. 1968. Infertility in inbred ewes. J. Reprod. Fert., 15, 277-282.

Dufour, J.J., Fahmy, M.H. 1975. Embryonic mortality and development during early pregnancy in three breeds of swine with purebred and crossbred litters. Can. J. Anim. Sci., 55, 9-15.

Dziuk, P.J. 1968. Effect of number of embryos and uterine space on embryo survival in the pig. J. Anim. Sci., 27, 673-676.

Edey, T.N. 1969. Prenatal mortality in sheep : a review. Anim. Breed. Abstr., 37, 173-190.

Epstein, H. 1969. Domestic animals of China. Commonwealth Agricultural Bureaux, Farnham Royal, Bucks., England

Fechheimer, N.S. 1981. Cytongentics in pig production. Pig News and Information, 2, 387-391.

Fenton, F.R., Bazer, F.W., Robison, O.W., Ulberg, L.C. 1970. Effect of quantity of uterus on uterine capacity in gilts. J. Anim. Sci., 31, 104-106.

Flint, A.P.F., Saunders, P.T.K., Ziecik, A.J. 1982. Blastocyst-endometrium interactions and their significance in embryonic mortality. Cole, D.J.A., Forcroft, G.R. Control of pig reproduction, 253-276. Butterworth Scientific, London.

Ford, C.E., Evans, E.P. 1973. Robertsonian translocations in mice: segretational irregularities in male heterozygotes and zygotic unbalance. Chromosomes today, 4, 387-397.

Glahn-Luft, B., Wabmuth, R. 1980. The Influence of the 1/20 translocation in sheep on the efficiency of reproduction. 31st annual meeting of the European association for animal production, Munich 1-4 Sept. 1980 (commission in sheep and goat production).

Hagen, D.R., Kephart, K.B. 1980. Reproduction in domestic and feral swine. I. Comparison of ovulation rate and litter size. Biol. Reprod. 22, 550-552.

Hanly, S. 1961. Prenatal mortality in farm animals. J. Reprod. Fert., 2, 182-194.

Hanrahan, J.P. 1980. Ovulation rate as the selection criterion for litter size in sheep. Proc. Aust. Soc. Anim. Prod., 13, 405-408.

Hanrahan, J.P. 1982. Selection for increased ovulation rate, litter size and embryo survival. 2nd World congress on genetics applied to livestock production, Madrid, 4-8 Oct. 1982. 5, 294-309. Editorial Garsi, Madrid.

Hanrahan, J.P., Quirke, J.F. 1976. Maternal performance of Finnish Landrace, Galway and Fingalway ewes. Anim. Prod., 22, 162 (abstr).

Hanrahan, J.P., Quirke, J.F. 1977. An egg transfer study of uterine capacity and embryo mortality in sheep. J. Anat., 124, 490 (abstr).

Hughes, P., Varley, M. 1980. Factors affecting embryonic and fetal mortality. "Reproduction in the Pig", 107-117, Butterworth & Co. London.

Hulot Francoise, 1969. Nouveau cas de fusion centrique chez la chevre domestique (Capra hircus L.). Ann. Genet. Sel. Anim., 1, 175-176.

Jensen, E.L., Smith, C., Baker, L.N., Cox, D.F. 1968. Quantitative studies on blood group and serum protein systems in pigs. II. Effects on production and reproduction. J. Anim. Sci, 27, 856-862.

Johansson, K. 1981. Some notes concerning the genetic possibilities of improving sow fertility. Livest. Prod. Sci., 8, 431-447.

Johnson, R.K. 1981. Crossbreeding in swine : experimental results. J. Anim. Sci., 52, 906-923.

Johnson, R.K., Omtvedt, I.T., Walters, L.E. 1978. Comparison of productivity and performance for two-breed and three-breed crosses in swine. J. Anim. Sci., 46, 69-82.

Johnson, R.K., Zimmerman, D.W., Kittok, R.J. 1984. Selection for components of reproduction in swine. Livest. Prod. Sci., 11, 541-558.

King, W.A., Gustavsson, I., Popescu, C.P., Linares, T. 1981. Gametic products transmitted by rep (13q-;14q+) translocation heterozygous pigs, and resulting embryonic loss. Hereditas, 95, 239-246.

King, R., Williams, I.H. 1984. The influence of ovulation rate on subsequent litter size in sows. Theriogenology, 21, 677-680.

Lahlou-Kassi, A., Marie, M. 1981. A note on ovulation rate and embryonic survival in D'man ewes. Anim. Prod., 32, 227-229.

Land, R.B. 1978. Genetic improvement of mammalian fertility : a review of opportunities. Anim. Reprod. Sci., 1, 109-135.

Larsen, W.A., McDonald, M.F. 1971. Reproductive differences between Border Leicester X Romney and Romney two-tooth ewes. Proc. N.Z. Soc. Anim. Prod., 31, 176-179.

Lauvergne, J.J. 1969. Progres des connaissances genetiques dur l'inter-sexualite associee a l'absence de cornes chez la chevre d'origine alpine. Ann. Genet. Sel. Anim., 1,403-412.

Lawson, R.A.S., Rowson, L.E.A. 1972. The influence of breed of ewe and offspring on litter size after egg transfer in sheep. J. Reprod. Fert., 28, 433-439.

Lax, J., Brown, G.H. 1968. The influence of maternal handicap, inbreeding and ewe's body weight at 15-16 months of age on reproduction rate in Australian Merinos. Aust. J. Agric. Res., 19, 433-442.

Legault, C., Caritez, J.C. 1983. L'experimentation sur le porc chinois en France. I. Performance de reproduction en race pure et en croisement. Genet. sel. Evol., 15, 225-240.

Legault, C., Gruand, J. 1976. Amelioration de la prolificite des truies par la creation d'une lignee hyperprolifique et l'usage de l'insemination artificielle : principe et resultats experimentaux preliminaires. 8eme journees de la Recherche porcine en France, Paris, 4-6 Fevr. 1976. 201-206, ITP. Paris.

Legault, C., Gruand, J. 1981. Effets additifs et non additifs des genes sur la precocite sexuelle, le taux d'ovulation et la mortalite embryonnaire chez la jeune truie. 13eme Journees de la Recherche porcine en France, Paris, 4-5 Fevr., 1981, 247-254, ITP, Paris.

Legault C., Gruand, J., Bolet, G. 1981. Resultats de l'utilisation en race pure et en croisement d'une lignee dite "hyperprolifique". 13eme Journees de la Recherche porcine en France, Paris, 4-5 Fevr. 1981, 255-260, ITP, Paris.

Long Susan, E. 1977. Cytogenetic examination of preimplantation blastocysts of ewes mated to rams heterozygous for the Massey I (tl) translocation. Cytogenet. Cell. Genet., 18, 82-89.

Long Susan, E., Williams, C.W. 1980. Frequency of chromosomal abnormalities in early embryos of the domestic sheep (Ovis aries). J. Reprod. Fert., 58, 197-201.

Lyngset, O. 1968. Embryonic mortality in the goat. VIe Cong. Intern. Reprod. Anim. Insem. Artif., Paris, 22-26 July 1968, 1, 439-441, INRA Paris.

Lutter, K. 1981. Untersuchungen zum Auftreten embryonaler und fetaler Feckelverluste und praktische Hinweise zu ihrer Verringerung. Mh. Vet. Med., 36, 6-11.

Lutter, K., Huhn, R., Huhn, V., Kaltofen, U., Lampe, B., Schneider, F. 1981. Untersuchungen zum pranatalen Fruchttod bei Jungsauen sowie zum Wachstum der Embryonen und Feten. Arch. exper. Vet. Med., Leipzig, 35, 687-695.

Marlowe, T.J., Smith, J.H. 1971. Early prenatal death loss in pigs. J. Anim. Sci., 33, 203 (abstr).

McFee, A.F., Banner, M.W. 1969. Inheritance of chromosome number in pigs. J. Reprod. Fert., 18, 9-14.

McFeely, R.M. 1967. Chromosome abnormalities in early embryos of the pig. J. Reprod. Fert., 13, 579-581.

Meyer, H.H., Clarke, J.N. 1982. Effect of ewe ovulation rate and uterine efficiency on breed and strain variation in litter size. Proc. N. Z. Soc. Anim. Prod. 42, 33-35.

Moore, N.W. 1968. The survival and development of fertilized eggs transferred between Border Leicester and Merino ewes. Aust. J. Agric. Res., 19, 295-302.

Moore, N.W., Eppleston, J. 1979. Embryo transfer in the Angora goat. Aust. J. Agric. Res., 30, 973-981.

Moore, N.W., Rowson, L.E.A. 1960. Egg transfer in sheep. Factors affecting the survival and development of transferred eggs. J. Reprod. Fert., 1, 332-349.

Moulick, S.K., Guha, H., Gupta, S., Mitra, D.K., Bhattacharya, S. 1966. Factors affecting multiple birth in Black Bengal goats. Ind. J. Vet. Sci. Anim Husb., 36, 154-163.

Nair, K.P., Raja, C.K.S.V. 1973. Studies on the gravid genitalia of goats. Ind. Vet. J., 50, 42-50.

Ollivier, L. 1982. Selection for prolificacy in the pig. Pig News and Information, 3, 383-388.

Ollivier, L., Bolet., G. 1981. La selection sur la prolificite chez le porc : resultats d'une experience de selection sur dix generations. 13eme Journees de la Recherche porcine en France, Paris 4-5 Fevr. 1981, 261-267, ITP, Paris.

Ollivier, L., Legault, C., 1967. L'influence directe du verrat sur la taille et la poids des portees obtenues par insemination artificielle. Ann. Zootech., 16, 247-254.

Ollivier, L., Sellier, P. 1982. Pig genetics : a review. Ann. Genet. Sel. Anim., 14, 481-544.

Padeh, B., Wysoki, M., Soller, M. 1971. Further studies on a Robertsonian translocation in the Saanen dairy goat. Cytogenetics, 10, 61-69.

Pen Jun Quian, P. 1982. A survey of some special chinese native goat breeds. 3d Intern. conference on goat production and disease, Tucson, 10-15 Jan. 1982, Dairy Goat Journal publishing Co., Scottsdale, U.S.A.

Perry, J.S., Rowlands, I.W. 1962. Early pregnancy in the pig. J. Reprod. Fert., 4, 175-188.

Polge, C., 1982. Embryo transplantation and preservation. In D.J.A. Cole, G.R. Foxcroft "Control of pig reproduction", 277-292, Butterworths, London.

Polge, C., Rowson, L.E.A., Chang, M.C. 1966. The effect of reducing the number of embryos during early stages of gestation on the maintanance of pregnancy in the pig. J. Reprod. Fert., 12, 395-397.

Pope, C.E., Christensen, R.K., Zimmerman-Pope V.A., Day, B.N., 1972. Effect of number of embryos on embryonic survival in recipient gilts. J. Anim. Sci., 35, 805-808.

Popescu, C.P. 1972. Mode de transmission d'une fusion centrique dans la descendance d'un bouc (Capra hircus L.) heterozygote. Ann. Genet. Sel. Anim., 4, 355-361.

Popescu, C.P., Bonneau, M., Tixier Michele, Bahri, I., Boscher Jeanine, 1984. Reciprocal translocations in pigs. J. Hered., 75, (in Press).

Popescu, C.P., Boscher Heanine, 1982. Cytogenetics of preimplantation embryos produced by pigs heterozghous for the reciprocal translocation (4q+ ; 14q-). Cytogenet. Cell. Genet., 34, 119-123.

Popescu, C.P., Tixier Michele, 1984. L'incidence des anomalies chromosomiques chez les animaux de ferme et leurs consequences economiques. Ann. Genet., 27, 69-72.

Quinlivan, T.D., Martin, C.A., Taylor, W.B., Cairney, I.M. 1966. Estimates of pre- and perinatal mortality in the New Zealand Romney Marsh ewe. J. Reprod. Fert., 11, 379-390.

Rao, V.H., Bhattacharva, N.K. 1980. Ovulation in Black Bengal nanny goats. J. Reprod. Fert., 58, 67-69.

Rasmusen, B.A., Hagen Karen L. 1973. The H blood-group system and reproduction in pigs. J. Anim. Sci., 37, 568-573.

Renard Christine, Bolet, G., Dando, P., Vaiman, M., 1985. Relations d'un marqueur genetique, le complexe majeur d'histocompatibilite, avec la prolificite des truies et la mortalite des porcelets. 17 e Journees de la recherche porcine en France, Paris, 30 Jan.- ler fevr. 1985, ITP-Paris (in press).

Restall, B.J., Brown, G.H., Blockey, M. de B., Cahill, L., Kearins, R. 1976. Assessment of reproductive wastage in sheep. I. Fertilization failure and early embryonic survival. Aust. J. Exp. Agric. Anim. Husb., 16, 329-335.

Restall, B.J., Griffiths, D.A., 1976. Assessment of reproductive wastage in sheep. 2. Interpretation of data concerning embryonic mortality. Aust. J. Exp. Agric. Anim. Husb., 16, 336-343.

Ricordeau, G. 1969. Surprolificite des genotypes sans cornes dans les races caprines Alpine Saanen, Alpine chamoisee et Poitevine. Ann. Genet. sel. anim., 1, 391-395.

Ricordeau, G. 1972. Observations sur les caracteres de reproduction des produits males et femelles issus d'un bouc porteur d'une "fusion centrique". Ann. Genet. Sel. Anim. 4, 593-598.

Ricordeau, G. 1982. Major genes in sheep and goats. 2nd World congress on genetics applied to livestock production, Madrid, 4-8 Oct. 1982. 6, 454-461, Editorial Garsi, Madrid.

Ricordeau, G., Razungles, J., Eychenne, F., Tchamitchian, L. 1976.
Performances de reproduction des brebis Berrichonnes due Cher,
Romanov et croisees. Ann. Genet. Sel. Anim., 8, 25-35.
Ricordeau, G., Sanchezy, G.F.F., 1981. Evolution de la frequence du
cornage dans quatre races caprines francaises. Ann. Genet. Sel.
Anim., 13, 353-362.
Ricordeau, G., Razungles, J., Lajous, D. 1982. Geritability of ovulation
rate and level of embryonic losses in Romanov breed. 2nd World
congress on genetics applied to livestock production Madrid, 4-8 Oct.
1982, 7, 591-595. Editorial Garsi, Madrid.
Riera, S. 1982. Reproductive efficiency and management in goats. 3rd Int.
Nat. conference on goat production and disease, Tucson, U.S.A., 10-15
Jan.,1982, 162-174. Dairy goat J. Publ. Co. Scottsdale, U.S.A.
Rombauts, P., Mazzari, G., du Mesnil du Buisson, F. 1982. Premier bilan
de l'experimentation sur le porc Chinois en France. 2. Estimation
des composantes de la prolificacite : taux d'ovulation et survie
foetale. 14eme Journees de la Recherche porcine en France, Paris,
3-4 fevr. 1982, 137-142, ITP, Paris.
Schinkel, P.G. 1955. Inheritance of birthcoats in a strain of Merino
sheep. Aust. J. Agric. Res., 6, 604-607.
Schworer, D., Blum, J.K., Sempach, M.L.P., Kristensen, B. 1984. Beziehungen
zwischen den Histokompabilitats Antigenen und diversen Fruchtbarkeit-
sparametern beim Schwein. Schweiz. Landw. Monatshefete, 8/9, 268-272.
Schworer, D., Morcel, P. 1984. Beziehung zwischen der Halothanreaktion und
der Fruchtbarkeit beim Schwein. Schweiz Landw. Monatshefte, 8/9,
260-267.
Schofield, A.M. 1972. Embryonic mortality. In D.J.A. Cole,"Pig Production"
367-384, Butterworths, London.
Scofield, A.M., Clegg, G.G., Lamming, G.E. 1974. Embryonic mortality and
uterine infection in the pig. J. Reprod. Fert., 36, 353-361.
Sellier, P. 1976. The basis of crossbreeding in pigs : a review. Livest.
Prod. Sci., 3, 203-226.
Sellier, P. 1982. Le choix de la lignee male du croisement terminal chez
le porc. 14e Journees de la Recherche porcine en France, Paris,
3-4 Fevr. 1982. 159-182, ITP, Paris.
Shelton, M. 1978. Reproduction and breeding of goats. J. Adiry Sci., 61,
994-1010.
Smith, J.H., Marlowe, T.J. 1971. A chromosomal analysis of 25 day-old
pig embryos. Cytogenetics, 10, 385-391.
Soller, M., Angel Huguette, 1964. Polledness and abnormal sex rations in
Saanen goats. J. Hered., 55, 139-142.
Soller, M., Kempenich, Ora. 1964. Polledness and litter size in Saanen
goats. J. Hered., 55, 301-304.
Soller, M., Wysoky, M., Padeh, B., 1966. A chromosomal abnormality in
phenotypically normal Saanen goats. Cytogenetics, 5, 88-93.
Swiestra, E.E., Dick, G.W. 1976. Influence of the boar and ejaculation
frequency on pregnancy rate and embryonic survival in swine.
J. Anim. Sci., 42, 455-460.
Thibault, C. 1959. Analyse de la fecondation de l'oeuf de truie apres
accouplement ou insemination artificielle. Ann Zootech., 8, (supp),
165-188.
Tomes, G.J., Newmann, R.B. 1984. Selection for litter size in pigs.
Anim. Prod. Aust., 15, 760.
Turner, Helena N. 1969. Genetic improvement of reproduction rate in sheep
Anim. Breed. Abstr., 37, 545.

Uzu, G. 1979. Influence du verrat sur les principaux parametres de la productivite du troupeau et sur la duree de gestation. Ann. Zootech., 28, 315-323.

Walker, S.K., Obst, J.M., Smith, D.H., Hall, G.P., Flavel, P.F., Ponzoni, R.W. 1981. A note on haemoglobin type and reproductive performance of ewes grazing oestrogenic pastures. Anim. Prod., 32, 223-224.

Webb, A.J., Jordan, C.H.C. 1979. Halothan sensitivity as a field test for stress-susceptibility in the pig. Anim. Prod., 26, 157-168.

Webel, S.K., Dziuk, P.J. 1974. Effect of stage of gestation and uterine space on prenatal survival in the pig. J. Anim. Sci., 38, 960-963.

Wentzel, D. 1982. Non-infectious abortion in Angora goats. 3rd Intern. conf. on goat production and disease, Tuscon, 10-15 Jan. 1982. 155-161, Dairy goat Journal publishing Co., Scottsdale, U.S.A.

White, D.H., Rizzoli, D.J., Cumming, I.A. 1981. Embryo survival in relation to number and site of ovulation in the ewe. Aust. J. Exp. Agric. Anim. Husb. 21, 32-38.

Willeke, H. 1981. Die Prufung der Eber auf Fruchtbarkeit, eine zuchterische und okonomische Notwendigkeit, Zuchthyg., 16, 171-175.

Willeke, H., Amler, K., Fishcer, K. 1984. Der Einflub des Halothanstatus der Sau auf deren Wurfgrobe. Zuchtungskunde, 56, 20-26.

Willeke, H., Richter, L. 1981. Der Einflub des Ebers auf die Wurfgrobe beim schwein. Zuchtungskunde, 52, 438-443.

Young, L.D., Johnson, R.K., Omtvedt, I.T. 1976. Reproductive performance of swine bred to produce purebred and two-breed cross litters. J. Anim. Sci., 42, 1133-1149.

Young, L.D., Johnson, R.K., Omtvedt, I.T. 1977. An analysis of the dependency structure between a gilt's prebreeding and reproductive traits. I. Ohenotypic and genetic correlation. J. Anim. Sci., 44, 557-564.

DISCUSSION

Wilmut:

The hypothesis of Bishop (1964) that embryonic loss is of
genetic origin and unavoidable is fatalistic. In mice 10
generations of selection increased embryo survival from 80 to
90% and it was shown that survival rate was determined by the
mother; similarly we should attempt to increase embryo
survival in farm animals.

Sreenan:

The cytogenetic studies reported by Gayerie (1982) suggest
a very low (3-5%) incidence of chromosome abnormality in cow
embryos supporting the view that a proportion of embryo loss
is avoidable.

Bouters:

Twinning rate is about 4% in dairy cows and 1% in heifers
although ovulation rate is similar for both groups. Uterine
capacity is probably the limitation in heifers yet you report
similar embryonic survival rates in bilateral and unilateral
twin pregnancies. Can you comment?

Sreenan:

I am unaware of the published data showing a similar
ovulation rate for heifers and cows. In our own laboratory
we have recorded a spontaneous twin ovulation rate of 1.4%
for 286 heifers which suggests a significantly lower ovulation
rate than for cows. Embryo transfer data shows clearly that
ovulation rate and not uterine capacity is the main cause of
low twinning rates in heifers.

Greve:

Philipsen has reported a higher proportion of spontaneous
bilateral (60%) than unilateral (40%) twin pregnancies. Can
you comment on this?

Sreenan:

   The literature in this area is very limited, somewhat
contradictory and is mainly confined to abattoir surveys.
The data I have presented show that following embryo transfer
a single uterine horn can support twin pregnancy with the
same frequency as both uterine horns (bilateral).   If the
difference you suggested exists in spontaneous twinning then
this would perhaps suggest a lower fertilisation rate following
unilateral twin ovulations.

Heap:

   The incidence of chromosomal abnormalities in cattle is
reported to be 3-5%.   What is the incidence in Meishan pigs
and to what extent might it be related to an increased uterine
horn size?

Bolet:

   The value of 3-5% for cattle is similar to that for pigs
and while the main effect is probably due to the sow, there
may also be foetal effects.   The problem of uterine capacity
is to find a way to measure it and then select for it.

MOLECULAR AND CELLULAR SIGNALLING AND EMBRYO SURVIVAL

R.B. Heap, V. Rider, F.B.P. Wooding, A.P.F. Flint

Agricultural and Food Research Council
Institute of Animal Physiology
Babraham, Cambridge CB2 4AT, U.K.

ABSTRACT

Bi-directional signals are transferred between embryo and mother during the time when embryo loss is high. The identity and function of embryonic signals differs in various farm animals. Oestrogens and/or proteins produced by the pre-implantation blastocyst play an important part in the prolongation of the life-span and function of the corpus luteum. Early pregnancy factor is first detected when embryos are in the oviduct, but its physiological role is uncertain. Binucleate cells differentiate in trophectoderm just prior to the onset of implantation and are directly involved in nidation and subsequent formation of the definitive placenta. In addition to these aspects of embryo and trophoblast development, progesterone is an indispensible signal from the mother and a recent study in mice shows that the physiological effects of progesterone depletion are first reflected in the arrest of embryo development during tubal life with subsequent embryo loss. Oxytocin of ovarian origin is involved in luteal regression, whereas maintenance of the corpus luteum is associated with the cessation of luteal oxytocin secretion, a loss of endometrial oxytocin receptors and a reduction in the pulsatile secretion of prostaglandin F-2α by the gravid uterus. It is suspected, but not yet proved in every instance, that failure to produce these bi-directional signals in sufficient quantity and at the right time may be one of the underlying causes of early embryonic mortality in farm animals.

INTRODUCTION

Maternal recognition of the presence of a conceptus and the adaptations associated with the establishment of pregnancy are achieved by different means in various species. Among those species studied in detail short-range signals of embryonic origin have been identified which act within the gravid uterus, and long-range signals have been shown to influence the activities of endocrine organs (anterior pituitary and ovary), the characteristics of the immune network including the draining lymph node, and as pregnancy progresses, the metabolic and cardiovascular features of the mother. In this paper we shall concentrate on the nature of these signals because it is suspected that the failure to produce them in sufficient quantity and at the right time underlies a substantial part of the high incidence of early embryonic mortality which occurs in all mammalian species that have been investigated (reviewed by Hanly, 1961; Boyd, 1965; Short, 1979). Estimates of embryo loss from counting the number of corpora lutea compared with the number of embryos implanted range from 25-40% (pig, Flint et al., 1982; ewe, Edey, 1979; cow, Ayalon, 1978) with most losses occurring during the first 25 to 30 days of pregnancy (Perry and Rowlands, 1962; Roche et al., 1981; Sreenan and Diskin, 1983). Factors such as high ambient temperature, severe undernutrition, disease and stress are known to increase embryo mortality above the basal level (Edey, 1979), but it is uncertain whether basal loss is due primarily to genetic or environmental components.

Since the establishment and maintenance of pregnancy consist of a series of complex endocrinological, physiological and immunological events, progress towards identifying the relative importance of individual factors which contribute to embryo loss has been slow. In humans, the frequency of genetic abnormalities in spontaneously aborted fetuses in the first trimester of pregnancy is high (Boue et al., 1975; Lauritsen, 1982; Biggers, 1969), though in domestic species they appear to account for less than 5% of embryo loss (Boyd, 1965). However, karyotypes in domestic animals have not been studied to the same extent as in man (Gustavsson, 1984).

During the last fifteen years evidence has accumulated in support of the concept that embryos secrete chemicals to signal their presence in the uterus (Psychoyos, 1966; Perry et al., 1973; Flint et al., 1979; Tyndale-Biscoe, 1979). The elaboration of this signal ensures the maintenance of corpora lutea (Bazer and Thatcher, 1977) and the provision of a local environment with an adequate supply of nutrients (Ford and Christenson, 1979). Survival of the early embryo seems to depend in the first place on intrinsic programmed development (Johnson, 1979) and subsequently on the ability to communicate with the mother (Bazer and First, 1983). If this hypothesis proves to be correct, the genotype of the embryo may be critical during early development since this will influence the quantity and quality of embryonic secretory products. This is evident from the elegant experiments of Barton et al. (1984) which demonstrated the importance of the paternal genome for normal development of mouse embryonic membranes. Future research regarding embryo loss should be directed towards defining further the contributions of the respective genomes in early development and whether these components can be manipulated by selective matings to improve embryo survival (Perry, 1960).

Tubal life and early embryonic mortality

The efffects of bi-directional signalling between embryo and mother are demonstrable from a very early stage of pregnancy. Studies of early pregnancy factor (EPF) in mice, women and pigs (mouse, Morton et al., 1976; women, Morton et al., 1977; pig, Morton et al., 1983; Koch, 1985), and of the transport of fertilised eggs and the retention of unfertilised eggs in the oviduct of mares (van Niekerk and Gernecke, 1966; Betteridge, Eaglesome and Flood, 1979) suggest that in some species the presence of the zygote is recognised by the maternal organism soon after conception. Questions remain, however, about the importance of the tubal environment for the development of the early embryo during cleavage stages, and its relevance for subsequent embryo survival.

The early mammalian embryo spends the first three or four days of its existence in the oviduct where it is surrounded by fluid which contains serum-derived compounds (Feigelson and Kay, 1972; Oliphant et al., 1978), including electrolytes (Roblero et al., 1976; Borland et al., 1977), energy substrates (Hamner and Fox, 1969; Hamner, 1973; Leese and Aldridge, 1979; Nieder and Corder, 1982), and specific products from secretory cells of the oviducal epithelium (Mastroianni et al., 1970, 1973; Shapiro et al., 1971; Sacco and Shivers, 1973; Noske and Daniel, 1974; Oliphant and Ross, 1982; Oliphant et al., 1984). The volume and composition of the oviduct fluid change with the endocrine status of the female (Hamner and Fox, 1969; Hamner, 1973; Mastroianni and Go, 1979). Maximum volume is found at oestrus (cow, Roberts et al., 1975; sow, Hunter, 1977; ewe, Restall, 1965; Perkins, 1974) and the minimum at the mid-luteal stage of the cycle and during pregnancy. Oestrogen treatment stimulates the rate of secretion in ovariectomized ewes while progesterone reduces the amount of fluid (McDonald and Bellve, 1969). It is also clear that ovarian steroids affect the rate of transport of embryos through the oviduct into the uterus (Blandau, 1969, 1973; Greenwald, 1976; Roblero and Garavayno, 1979; Singh et al., 1983), and that premature entry of embryos into the uterus is detrimental to their survival in laboratory animals, as shown by Chang in 1950. The effect of steroids on tubal transport differs according to the species, the dose of steroid administered and the time of treatment in relation to ovulation (Greenwald, 1967; Blandau, 1973; Chang,

1976), but in general, oestrogen treatment retards (Burdick and Pincus, 1935; Humphry, 1976) whereas progesterone accelerates transport into the uterus (Day and Polge, 1968; Hunter, 1980).

Whereas the hormonal regulation of the tubal environment and tubal transport is well established, it does not follow that the environment regulates embryo development. Mouse (and human) embryos develop successfully in a defined medium in culture (Brinster, 1963; see Beier and Lindner, 1983). Fertilised eggs can develop normally after transfer to the oviduct of ovariectomized ewes (Ryan and Moore, 1983). Pregnancy can occur in rabbits after the surgical implantation of an ovary in the uterine wall thus circumventing exposure of the fertilised egg to the tubal environment (Estes' operation, Adams, 1979). These, and other studies, appear to support the notion that the zygote proceeds through the early cleavage divisions without any specific requirements from a tubal environment, and any perturbation of the environment at this phase of development is unlikely to interfere with subsequent survival. However, it is notable that in several studies quoted above the success rate of survival is low, and even with mouse embryos, which can be cultured in vitro with a high degree of success, rates of metabolism decline during culture (Surani and Fishel, 1980) implying that an important ingredient is deficient, or an embryotoxic compound or inhibitor is produced. Furthermore, pig embryos retained in the oviduct by ligation at the uterotubal junction do not develop into blastocysts (Murray et al., 1971), whereas those restricted to the ampullary region of the oviduct will progress to the early blastocyst stage at a slower rate than normal (Pope and Day, 1972). Embryos restricted to the oviduct by ligation in many other species will blastulate (mouse, Kirby et al., 1962; Whittingham, 1968; Weitlauf, 1971; rabbit, Adams, 1958, 1973; rat, Alden, 1942; ewe, Wintenberger-Torres, 1956) but their subsequent developmental capacity is impaired. Oviduct ligation may produce an abnormal oviducal environment and therefore provides little information about the normal physiological conditions that pertain in early pregnancy, so that alternative procedures are required to determine whether the tubal phase of early embryonic development is regulated in any way by maternal factors.

We have prepared an anti-progesterone monoclonal antibody to investigate this question in mice, and the results show that early embryonic development is arrested and embryonic mortality greatly increased as a result of passive immunisation. Nulliparous mice were injected at 32 h after mating with a single intraperitoneal dose of ascites fluid containing anti-progesterone immunoglobulin G (IgG). At this stage of pregnancy the concentration of progesterone in circulating plasma is similar to that in non-pregnant females (< 10 ng/ml). Pregnancy was completely blocked in BALB/c mice at a dose calculated to bind greater than 90% of steroid in circulation (Wright et al., 1982). Similar results have been obtained in another inbred strain (CBA) of mice though the dose required was higher (13 nmol) than that in BALB/c (Table 1).

TABLE 1   Anti-implantation effect of progesterone monoclonal antibody in two inbred strains (BALB/c, CBA) of nulliparous mice.   A single injection of ascites fluid containing anti-progesterone immunoglobulin G (IgG) (treated) or non-immune ascites fluid (McPc 1748) (control) in 200 μl 0.9% NaCl was given 32 h post coitum (p.c.).   Autopsies were carried out at day 10 p.c. (day of plug = day 1 pregnancy) (from Wang et al., 1984).

Strain	Dose (nmol IgG)	No. pregnant/ no. treated	Implantation sites	
			Total no.	No./pregnant animal (Mean ± s.e.m.)
BALB/c	0	4/6	32	8.0 ± 2.4
	8	0/5	0	0
	13	0/5	0	0
CBA	0	8/8	56	7.0 ± 1.1
	8	3/8	23	7.7 ± 0.9
	0	6/6	44	7.3 ± 0.9
	13	0/6	0	0

Further investigation into the nature of the anti-implantation effect revealed that one of the first actions of antibody treatment was an arrest of early embryonic development so that the number progressing to the blastocyst stage in treated females was significantly reduced (Table 2).

The concentration of progesterone in the plasma of treated females at day 10 after mating was about 4-fold higher than that in control pregnant mice.   This measurement of total progesterone concentration in plasma includes steroid that is bound to antibody as well as that which is free or unbound.   Current studies indicate that high concentrations of plasma progesterone after antibody treatment are due to high-affinity binding of progesterone to antibody, thereby preventing the interaction of steroid with target cell receptors. This interpretation is consistent with the finding that the endometrium of antibody-treated females is insensitive to a decidual stimulus at the normal time (Rider et al., 1985). Furthermore, the effects of antibody on the early embryo are reversed by the administration of excess progesterone.

TABLE 2   Passive immunization with progesterone monoclonal antibody arrests development of the early mouse embryo. Animals were given a single intraperitoneal injection of ascites fluid containing 9.5 nmol anti-progesterone immunoglobulin G. Control females were given an equivalent volume (250 μl) of 0.9% NaCl. Embryos were flushed from the reproductive tract at day 4 post coitum with phosphate-buffered saline. Only those embryos which had started to cavitate were scored as blastocysts.

Strain	Treatment	No. of embryos		
		Total recovered (mean per female)	Developmental stage	
			Precavitation (%)	Blastocyst (%)
BALB/c	Antibody	45 (5.0)	38 (84)	7 (16)
	Control	58 (6.4)	4 (7)	54 (93)
CBA	Antibody	26 (3.7)	25 (96)	1 (4)
	Control	42 (6.0)	16 (38)	26 (62)

These results indicate that in the mouse a low but critical concentration of progesterone has an important role during the earliest stages of embryo development in vivo since elimination of this maternal signal arrests embryonic development. In this regard it is notable that the concentration of progesterone in oviducal fluid increases to values greater than those in plasma after ovulation in the rhesus monkey (Wu et al., 1977) and rabbit (Richardson and Oliphant, 1981) probably as a result of active secretion (Mastroianni and Go, 1979). Active secretion is stimulated by progesterone itself and is associated with the release of secretory granules from the oviducal epithelium (Greenwald, 1958; Brower and Anderson, 1969; Lambert et al., 1973; Stone et al., 1980). There is little evidence that progesterone has a direct effect on embryo development in vitro, and the ability of cultured mouse embryos to develop normally in a defined medium alluded to previously indicates that an action of progesterone on the oviduct in vivo may be to eliminate an inhibitor of embryo cleavage rather than to enhance the secretion of a growth factor. High molecular weight glycoproteins in the oviduct lumen at oestrus in sheep (Sutton et al., 1983) and during pseudopregnancy or after progesterone administration in rabbits (Stone et al., 1980) require further consideration in this respect.

SIGNALS DURING UTERINE LIFE

Maternal progesterone secretion is indispensible for the maintenance of pregnancy in all mammals studied in detail. Embryo survival in utero depends on the prolongation of the life-span and function of the corpus luteum in all farm animals (cow, pig, goat and sheep) since ovariectomy at most stages of gestation results in pregnancy failure in cows, pigs and goats, though in sheep the operation can be tolerated if it is performed after about day 50 of pregnancy. The classical studies of Rowson and Moor in the 1960s showed that in sheep the mother recognises the presence of an embryo in the uterus at about day 12-13 p.c. since the removal of embryos prior to this time results in the regression of the corpus luteum, but luteal maintenance occurs if they are removed at day 12 or later (Moor, 1968). A comparable stage of recognition has been discovered in cows, pigs and mares, and subsequently interest has focused on the nature of the signals produced by the embryo by which the corpus luteum is maintained. The signal is unlikely to be due to cell-cell contact because the time of maternal recognition of pregnancy (Short, 1969) antedates attachment and implantation (Table 3). It is improbable that neural pathways are involved since local interconnections between a uterine horn and its adjacent ovary have not been identified. The signal is probably not chorionic gonadotrophin (CG) since there is little proof that CG is produced by early trophoblast in these species.

TABLE 3   A comparison of the times of maternal recognition of pregnancy, attachment and parturition in farm animals (days after mating).

Species	Day by which embryo must be present for luteal maintenance	Definitive attachment	Parturition
Ewe	12-13	16	147
Cow	16-17	18-22	282
Sow	11-12	18	114
Mare	14-16	36-38	340

After Findlay (1983)

Chemical nature of embryonic signals

For the purpose of this review we shall concentrate on two types of chemicals that are produced by the early embryo which are involved in the obligatory event of prolongation of luteal maintenance, namely, steroids and proteins.

Oestrogens are luteotrophic in the pig (Kidder et al., 1955; Gardner et al., 1963) and are produced by the preimplantation pig blastocyst at the time of maternal recognition of pregnancy (Perry et al., 1973; Gadsby et al., 1980; Heap et al., 1981). Elongating blastocysts recovered from the uterus shortly after the onset of elongation at day 11 p.c. contain oestrone and oestradiol-17β; they possess many of the enzymes in the steroid biosynthetic chain including aromatase for oestrogen synthesis; and they produce oestrone and oestradiol-17β when incubated with radioactive precursors in vitro in the proportion of 9.7:1. Greater concentrations of oestradiol-17β (20 pg/ml) are detectable in utero-ovarian vein plasma of pregnant gilts between days 12 and 17 than in non-pregnant gilts (5 pg/ml) at the same days of the oestrous cycle (Moeljono et al., 1977), though subsequently oestrone sulphate is quantitatively more important in circulation. This conjugated oestrogen arises from sulphotransferase conversion of oestrogens within the gravid uterus and there is a peak concentration in peripheral circulation at about day 25-30 of gestation, followed by a decline at days 40-60 p.c. and a second peak prior to parturition (Robertson and King, 1974; Saba and Hattersley, 1981). Measurements of the total amount of oestrone, oestradiol-17β, oestriol, oestrone sulphate, oestradiol-17β-3-monosulphate and oestriol-3- monosulphate show that values are highest at day 12 p.c. and they then decline (Bazer et al., 1982). This finding may mean that blastocyst oestrogen production in the pig is associated with the important transition from the spherical to early filamentous stages of development when the blastocyst undergoes rapid remodelling and elongation (Geisert et al., 1982).

The only other species in which comparable aromatase activity in preimplantation embryonic tissue has been detected so far is the horse, but the time when this activity is most prominent occurs a few days before the onset of implantation at day 36 p.c. rather than around the time of the maternal recognition of pregnancy at day 15-17 p.c. (Fig. 1; Heap et al., 1982).

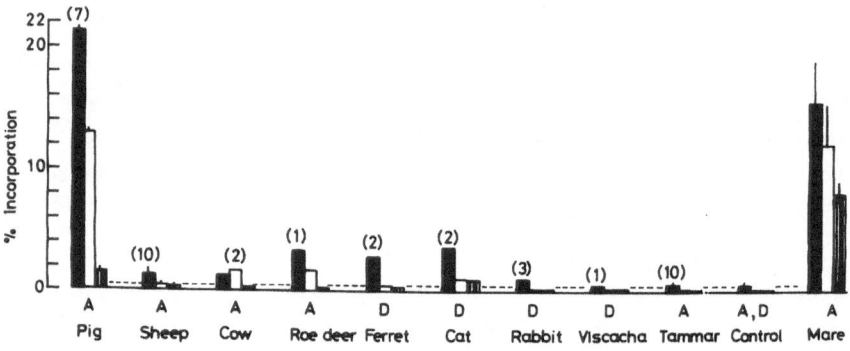

Fig. 1   Oestrogen synthesis by trophoblast tissue from various animals about the time of implantation or maternal recognition of pregnancy. Values show percentage incorporation of radio-labelled androstenedione (A) or dehydroepiandrosterone (D) into total phenolic compounds (black bar). Values in absence of tissue shown by horizontal broken line (for further details see Heap et al., 1982; reproduced by permission of the Journal of Reproduction and Fertility).

Proteins have been implicated as the embryonic signal involved in prolongation of luteal function in sheep (Rowson and Moor, 1967; Moor,

1968; Godkin et al., 1978; Martal et al., 1979; Ellinwood et al., 1979). Recent studies from Florida have shown the sheep conceptus produces two major secretory polypeptides between days 13 and 23 of pregnancy when cultured in vitro (Godkin et al., 1982; Masters et al., 1982). One of these, ovine trophoblast protein-1 (oTP-1) is a low molecular weight, acidic polypeptide ($M_r \backsim 17000$-21000, pI $\backsim 5.5$) and is produced only up to day 21 of pregnancy and not beyond. This protein has many of the characteristics required of an embryonic signal involved in maternal recognition since it is the only detectable product released by conceptuses cultured in vitro, it is synthesized during the critical phase of prolongation of luteal function, it accumulates in the uterine epithelium of day-16-pregnant ewes, and it binds to endometrial receptors (but is not detectable in maternal circulation). Moreover, proteins released by day 15-16 conceptuses (or purified oTP-1) prolonged luteal maintenance when introduced into the uterine lumen of cyclic ewes (Godkin et al., 1984). Evidence for conceptus secretory proteins with luteotrophic properties has also been reported in the cow (Knickerbocker et al., 1984).

The identification of oTP-1 as a potential anti-luteolytic or luteotrophic agent from the early conceptus confirms the original findings of Rowson and Moor (1967), and future work will undoubtedly be directed to answering the question of whether this is the only agent that is produced at this stage. Clearly, early trophoblast has the capacity to synthesize many compounds and other studies in farm and laboratory animals show that putative molecular signals including histamine (Dey et al., 1979a; Dey et al., 1979b) and prostaglandins (Kennedy,1980) are also produced about the time of implantation, another stage in the complex adaptation of the mother to the presence of an embryo in utero.

Mechanism of action

Pharriss and Wyngarden (1969) reported that PGF-2α is luteolytic in rats, and subsequent work showed this to be true in many, though not all mammalian species studied in detail (Horton and Poyser, 1976). PGF-2α is luteolytic in farm animals, though its efficacy varies according to the stage of the reproductive cycle when it is administered. The hypothesis that PGF-2α is an endogenous luteolysin produced in the endometrium, the primary site of synthesis in the sheep uterus, and that it causes the demise of the corpus luteum during the oestrous cycle is persuasively supported by the results of experiments in farm animals (sheep, McCracken et al., 1972; McCracken et al., 1981: pig, Moeljono et al., 1976; Moeljono et al., 1977; Bazer et al., 1982: cow, Nancarrow et al., 1973; Thatcher et al., 1984: goat, Maule Walker, 1983). Controversy continues, however, about whether or not PGF-2α is the sole uterine luteolysin (Watson and Maule Walker, 1977; Harrison, 1979; Hansel and Fortune, 1978).

A current view of the way by which the life-span of the corpus luteum is prolonged during pregnancy is that the molecular signal produced by the embryo at this stage inhibits (or reduces) the secretion of high-amplitude pulses of PGF-2α into circulation, thereby protecting the corpus luteum from exposure to a uterine luteolysin. Many explanations have been advanced to explain how oestrogens and proteins inhibit luteolysin release. They include the idea of oestrogen-dependent redirection of the secretion of PGF-2α into the uterine lumen, as in non-pregnant pigs treated with exogenous oestrogen, and in normal pregnant gilts which contain much higher concentrations of PGF in uterine flushings than are found in non-pregnant gilts (Bazer and Thatcher, 1977; Bazer et al., 1982). Uterine sequestration of PGF-2α has also been reported in sheep associated with increased concentrations (or binding) in the endometrium or uterine lumen during pregnancy (Harrison et al., 1976; Ellinwood, 1978; Ellinwood et al., 1979; Findlay et al., 1981). McCracken et al. (1981) have found that oestradiol-17β and progesterone control the luteolytic release of PGF-2α from the uterus by regulating the availability of oxytocin receptors in the endometrium. It has been proposed that the effect of the early embryo is to diminish the concentration of receptors for oxytocin and hence the secretion of PGF-2α by the endometrium, in addition to reducing the endometrial secretion of large pulses of PGF-2α (McCracken et al., 1984). The question of whether the rate of uterine secretion of PGF-2α is reduced during early pregnancy in sheep is disputed by some workers who consider that the increased production of PGE-2 of embryonic and/or endometrial origin may be responsible for protecting the corpus luteum against the actions of PGF-2α both in vivo and in vitro (Silvia et al., 1984; Ottobre et al., 1984).

Molecular signalling from embryo to mother in most of the examples we have considered so far consists of short-range negative feedback effects whereby chemicals elaborated by the embryo act locally on the endometrium to suppress the release of a uterine luteolysin. An important example of a positive feedback loop which is also interrupted during pregnancy has come to light recently from studies in the sheep. In the cyclic animal the pulsatile release of PGF-2α from the uterus may be related to the action of oxytocin of ovarian origin (Flint and Sheldrick, 1983; Sheldrick and Flint, 1983). Episodes of secretion of PGF-2α occur simultaneously with those of oxytocin during luteal regression and it has been suggested that this may reflect the operation of a positive feedback loop of the form illustrated in Figure 2 (Flint and Sheldrick, 1983). Consistent with these findings is the effect of immunisation against oxytocin, which reduces the effective concentrations of the hormone, leading to the prolongation of luteal function (Sheldrick et al., 1980; Schramm et al., 1983). Other studies have demonstrated that the corpus luteum produces oxytocin and its associated neurophysin (Wathes and Swann, 1982; Flint and Sheldrick, 1982; Watkins et al., 1983). Luteal concentrations of oxytocin decline dramatically after day 15 of gestation, along with peripheral circulating concentrations (Sheldrick and Flint, 1983a,b) and they also decline after the administration of cloprostenol (a synthetic analogue of PGF-2α) in association with luteal regression (Flint and Sheldrick, 1982; 1983). Luteal oxytocin concentrations also fall after hysterectomy (Sheldrick and Flint, 1983b), but in this case the corpus luteum is maintained because the uterine secretion of PGF-2α is lacking. Thus, the pulsatile release of endogenous PGF-2α is considered to induce the loss of luteal oxytocin, while the reduction in endometrial oxytocin receptors in early pregnancy serves to eliminate the feedback effect of oxytocin on uterine release of PGF-2α (Flint and Sheldrick, 1982, 1983; McCracken et al., 1984), thereby interrupting a positive feedback loop essential to the luteolytic mechanism. The effect of the embryo on endometrial cell receptors is probably of considerable duration since the corpus luteum persists for 20 to 40 days after the removal of the embryo from the uterus on day 21 after mating (McCracken et al., 1984). It remains to be proved whether progesterone itself or putative anti-luteolytic compounds such as oTP-1 reduce the concentration and/or affinity of oxytocin receptors for the ligand thereby blocking the positive feedback loop (Fig. 2). It should be noted that oxytocin has also been reported in the corpus luteum of the cow (Wathes et al., 1983; Dubois et al., 1983; Fields et al., 1983).

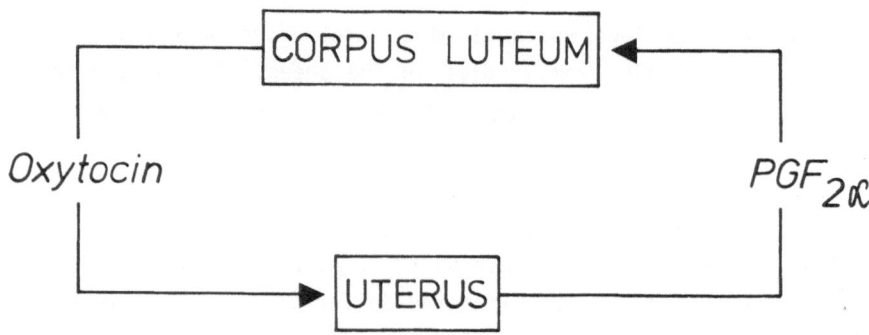

Fig. 2   Regulation of the corpus luteum and the role of a positive feedback between ovary and uterus (Flint and Sheldrick, 1983).

Positive effects may also occur in other ways to affect gonadotrophin and/or progesterone secretion. Thus, in pigs there is a dramatic rise in luteal LH receptor concentration between days 20 and 30 of gestation which can be mimicked by oestrogen administration to hysterectomised gilts, indicating that rising oestrogen production by the gravid uterus acts either directly on luteal receptor synthesis or indirectly through raised pituitary gonadotrophin secretion (Garverick et al., 1982).

OTHER MATERNAL ADAPTATIONS

The prolongation of the corpus luteum is clearly an essential event in the establishment of pregnancy in farm animals as we have previously outlined, but this event is accompanied by other maternal adaptations to the presence of a conceptus. Experiments in sheep showed that myometrial contractility and vasodilatory response to the purinergic agent, adenosine, were reduced significantly by the presence of a conceptus before the time of definitive attachment (Fleet and Heap, 1982). In pigs, uterine blood flow increases 2- to 4-fold from day 11 to days 12 and 13 of pregnancy probably reflecting the vasodilatory effects of oestrogens produced by the conceptus at this time (Ford and Christenson, 1979; Ford et al., 1982). Uterine blood flow increases at day 16 in the gravid, but not in the non-gravid, horn of cows (Ford et al., 1979) and this may be related to a conceptus production of a vasodilatory agent such as PGE-2 (Thatcher et al., 1984). Trophoblast-derived products have also been proposed to abrogate the maternal immune response to the presence of an embryonic allograft and these aspects have been reviewed elsewhere (Heap et al., 1983). So far as embryonic mortality is concerned, there is an urgent need for an experimental model in farm animals in which a distinction can be drawn between adaptations that are obligatory and those that are permissive in respect of embryo survival and implantation. One model involves the influence of raised ambient temperatures which decreases conception rates and increases embryonic mortality in several species including sheep, cows and pigs. However, a recent study of heat stress in gilts during the phase of maternal recognition of pregnancy failed to identify any significant change in the steroidogenic capabilities of embryos, or in uterine function (Wettemann et al., 1984), confirming the need for better models to improve our understanding of embryo loss at this stage of gestation.

CELLULAR SIGNALS AT IMPLANTATION

A significant proportion of early embryo loss is considered to occur about the time of implantation. Knowledge of the cellular events at implantation in farm animals has increased recently with detailed studies of cell-cell interactions at the feto-maternal interface (Guillomot et al., 1981, 1982a,b; King et al., 1980, 1981; Leiser, 1975; Wathes and Wooding, 1981; Wooding et al., 1982; Wooding, 1984). It is clear that in ruminants implantation is associated with the onset of fusion of fetal epithelial cells (trophectoderm) with the uterine epithelium (Wooding, 1984). Mechanisms may exist which regulate the rate of migration of these cells from one tissue to another. In contrast, the pig is a species with a non-invasive form of placentation, and the nature of cell-cell interactions is very different from that in ruminants. Mechanisms exist in this species which either maintain the integrity of the feto-maternal junction or prevent the expression of inherent trophoblast invasiveness/fusion, or both (Samuel and Perry, 1972). Examination of some of these events will show that our knowledge of these mechanisms is increasing, though it is inadequate to explain causes of implantation failure and embryo loss.

Implantation occurs late in pigs (day 12 to 13 p.c.) and ruminants (day 16 to 20 p.c., Table 3) by comparison with laboratory animals and man (Schlafke and Enders, 1975; Wimsatt, 1975), and only after a considerable elongation of the initially spherical conceptus has occurred (Wintenberger-Torrès and Fléchon, 1974). Implantation may be defined as the immobilisation of the conceptus in the uterus followed by cellular interactions with the uterine epithelium. The initial immobilisation of the relatively tiny conceptus in the enormous uterus takes place at a preferred site(s), the maternal caruncles in sheep, cows and goats and the mesometrial surface in the pig. In ruminants implantation frequently occurs at a site just proximal to the body of the uterus with usually one conceptus per horn (Lee et al., 1977). In pigs the conceptuses are spaced equally along the uterine horns. In both cases the positioning is considered to be caused by differential myometrial contractions with possibly some differential uterine growth (Wimsatt, 1975).

In the cow and sheep this positioning is stabilised by cellular sprouts from the conceptus which anchor into the necks of the uterine glands (Fig. 3a) (Wooding et al., 1982). From this fixed point the oval conceptus then elongates in both directions. The pig conceptus has no such sprouts but the complex foldings of the uterine wall may provide anchorage for the extreme elongation by cellular remodelling which converts the original sphere of 9 mm diameter into a long tubular pre-implantation blastocyst 100 cm in length (Geisert et al., 1982).

Once the conceptus has established a position in the uterus, the microvilli on the surface of the trophectodermal epithelium of the conceptus shorten and disappear as the fetal cells come into ever closer contact with the uterine epithelial microvilli. The two surfaces gradually fold to produce an interdigitation of fetal and maternal microvilli (Fig. 3b). In the pig subsequent growth of the placenta consists of the proliferation of this feto-maternal interface with no great modification of the apposed cellular layers.

In ruminants there is a considerable modification of the uterine epithelium by migration of trophectodermal binucleate cells (BNC) (Wooding, 1982). Binucleate cells form within the trophectoderm from uninucleate cells by nuclear division without a subsequent cellular division. They differentiate by synthesising numerous small dense granules and then migrate to the surface of the trophectoderm. Fusion between a mature binucleate cell and the uterine epithelial cell to which it is apposed produces a feto-maternal hybrid trinucleate cell (Fig. 3c) (Wooding, 1984). This then releases the granules derived from the

(a)   Immobilisation and elongation

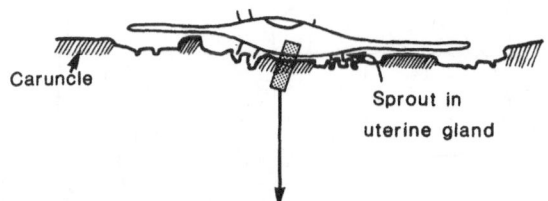

(b)   Cellular interaction and apposition

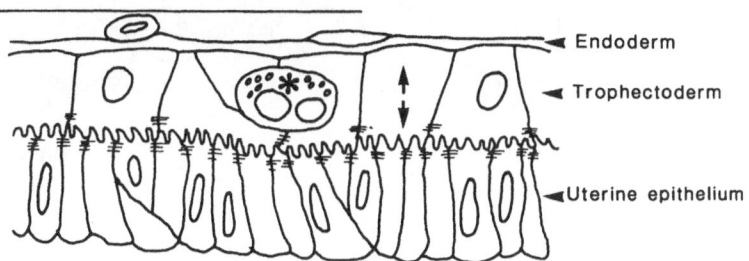

(c)   Binucleate cell migration

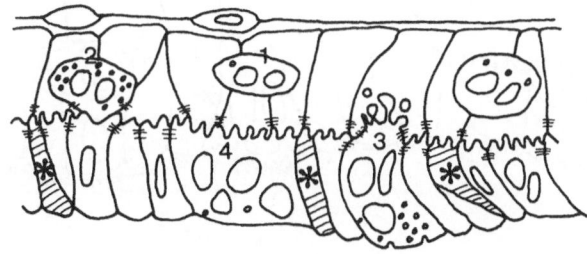

Fig. 3   The onset of implantation in ruminants. (a)Immobilis-
ation of conceptus by trophoblast sprouts. (b) Cellular interaction with caruncular epithelium and development of binucleate
cells, *.   (c)   Loss of uterine epithelium by migration and
fusion (1-4) of fetal binucleate cells with some uterine
epithelial cells and death of others, *.

(a)

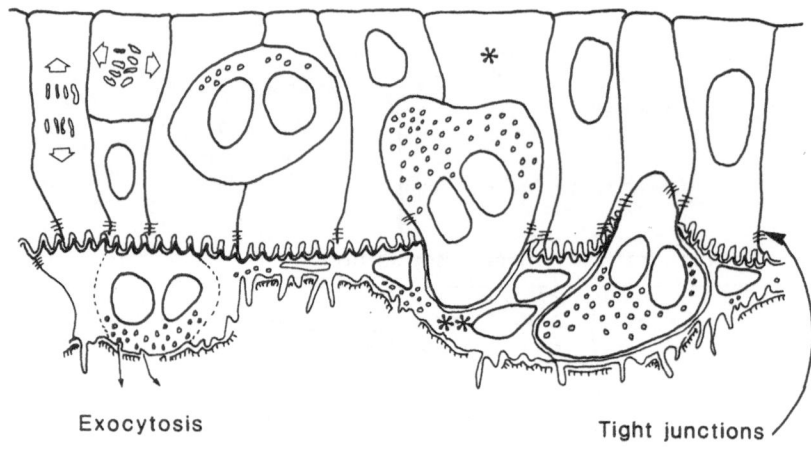

Exocytosis   Tight junctions

(b)

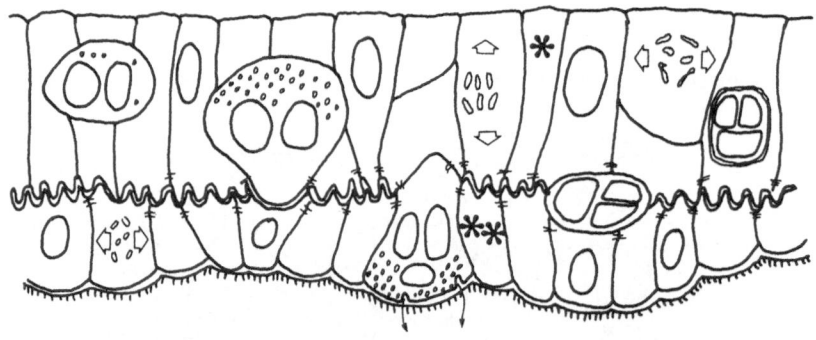

Fig. 4   Formation of definitive placenta. (a) Sheep and goat
with huge growth in area of trophectoderm (*) by cell division
and of the syncytium (**) by binucleate cell migration; (b) cow
and deer with huge growth in area of both trophectoderm (*) and
uterine epithelium (**) by cell division. Binucleate cell
migration produces transient trinucleate cells throughout
pregnancy.

binucleate cell to the maternal connective tissue by exocytosis. The first fusions take place on the caruncles, the specialised areas in the ruminant uterus destined to become the cotyledons of the mature placenta, but surface features that facilitate cell-cell interactions at these points are not known (but see Ricketts, Scott and Bullock, 1984). At this early stage there are far more BNC in the trophectoderm apposed to the caruncles than elsewhere, and as pregnancy progresses the BNC continue to mature, migrate and fuse with pre-existing trinucleate cells or other uterine epithelial cells. With coincident death of a number of the residual epithelial cells this produces a feto-maternal hybrid syncytium of the same area as the cellular uterine epithelim it has replaced (Fig. 3c). This syncytium is maintained and extended solely by binucleate cell migration during the subsequent enormous growth in area of the feto-maternal interface layers during the development of the definitive placental cotyledons in sheep and goats (Fig. 4a). In the cow and deer the caruncular syncytium is displaced by a regrowth of uterine epithelial cells, and the continual BNC migration produces trinucleate cells which release their granules and die (Fig. 4b) (Wooding and Wathes, 1980; Wooding, 1982).

In ruminants the number of BNC in the trophectoderm represent 15-20% of their total population throughout pregnancy. There is a very rapid initiation of production one or two days before implantation and a decrease in the last four or five days prior to parturition (Wooding, 1983). One common result of all BNC migration and fusion is delivery of the contents of the fetal granules to the maternal circulation, and in the sheep the granules have been shown to contain the hormone ovine placental lactogen (oPL) and other constituents (Martal et al., 1977; Watkins and Reddy, 1980; Wooding, 1981). The role of BNC migration in delivering a protein hormone to the maternal circulation is paralleled in another domestic animal, the horse. At implantation in the mare fetal trophectodermal binucleate cells migrate into the maternal uterine connective tissue and produce equine chorionic gonadotrophin, or PMSG, before the BNC are killed by maternal lymphocytes (Allen, 1982; Allen et al., 1973; Hamilton et al., 1973). In the ruminant the fusion of BNC with maternal tissue with consequent granule release may provide protection of the fetal allograft against cytotoxic maternal lymphocytes. There is also evidence that the BNC-derived oPL may play a direct role in enhancing fetal growth and maturation. A better understanding of the factors controlling the BNC production and migration could clearly play a very significant role in any strategy designed to improve ruminant reproductive efficiency.

ACKNOWLEDGEMENTS

The authors thank Mr M. Bacon, Miss M. Hamon and Mrs J. Tickner for help with preparing the manuscript and illustrations.

REFERENCES

Adams, C.E. 1958. Egg development in the rabbit: the influence of post coital ligation of the uterine tube and of ovariectomy. J. Endocr., 16, 283-293.

Adams, C.E. 1973. The development of rabbit eggs in the ligated oviduct and their viability after re-transfer to recipient rabbits. J. Embryol. exp. Morphol., 29, 133-144.

Adams, C.E. 1979. Consequences of accelerated ovum transport, including a re-evaluation of Estes' operation. J. Reprod. Fert., 55, 239-246.

Alden, R.H. 1942. Aspects of the egg-ovary-oviduct relationship in the albino rat. II. Egg development within the oviduct. J. exp. Zool., 90, 171-177.

Allen, W.R. 1982. Immunological aspects of the endometrial cup reaction and the effect of xenogeneic pregnancy in horses and donkeys. J. Reprod. Fert., Suppl. 31, 57-94.

Allen, W.R., Hamilton, D.W. and Moor, R.M. 1973. Origin of equine endometrial cups. II. Invasion of the endometrium by trophoblast. Am. J. Anat., 177, 485-502.

Ayalon, N. 1978. A review of embryonic mortality in cattle. J. Reprod. Fert., 54, 483-493.

Barton, S.C., Surani, M.A.H. and Norris, M.L. 1984. Role of paternal and maternal genomes in mouse development. Nature (Lond.), 311, 374-376.

Bazer, F.W. and First, N.L. 1983. Pregnancy and parturition. J. Anim. Sci., 572, 425-460.

Bazer, F.W. and Thatcher, W.W. 1977. Theory of maternal recognition of pregnancy in swine based on estrogen controlled endocrine versus exocrine secretion of prostaglandin $F_{2\alpha}$ by the uterine endometrium. Prostaglandins, 14, 397-401.

Bazer, F.W., Geisert, R.D., Thatcher, W.W. and Roberts, R.M. 1982. The establishment and maintenance of pregnancy. In "Control of Pig Reproduction" (Ed. D.S.A. Cole and G.R. Foxcroft). (Butterworth, London). pp. 227-252.

Beier, H.M. and Lindner, H.R. (Eds.) 1983. "Fertilization of the Human Egg In Vitro - Biological Basis and Clinical Application" (Springer Verlag, Berlin).

Betteridge, K.L., Eaglesome, M.D. and Flood, P.F. 1979. Embryo transport through the mare's oviduct depends upon cleavage and is independent of the ipsilateral corpus luteum. J. Reprod. Fert., Suppl. 27, 387-394.

Biggers, J.D. 1969. Problems concerning the uterine causes of embryonic death with special reference to the effects of ageing of the uterus. J. Reprod. Fert., Suppl. 8, 27-73.

Blandau, R.J. 1969. Gamete transport - comparative aspects. In "The Mammalian Oviduct" (Ed. E.S.E. Hafez and R.J. Blandau). (Univ. of Chicago Press, Chicago). pp. 129-162.

Blandau, R.J. 1973. Gamete transport in the female mammal. In "Handbook of Physiology" Endocrinology, section 7, Vol. II. (Ed. R.O. Greep). (American Physiology Society, Washington D.C.). pp. 153-163.

Borland, R.M., Hazra, S., Biggers, J.D. and Lechene, C.P. 1977. The elemental composition of the gametes and preimplantation embryo during the initiation of pregnancy. Biol. Reprod., 16, 147-157.

Boué, J., Boué, A. and Lazer, P. 1975. Retrospective and prospective epidemiological studies of 1500 karyotyped spontaneous human abortions. Teratology, 12, 11-26.

Boyd., H. 1965. Embryonic death in cattle, sheep and pigs. Vet. Bull., 35, 251-266.

Brinster, R.L. 1963. A method for in vitro cultivation of mouse ova from two-cell to blastocyst. Expl. Cell Res., 32, 205-208.

Brower, L.K. and Anderson, E. 1969. Cytological events associated with the secretory process in the genital tract of the rabbit (Oryctolagus cuniculus). Biol. Reprod., 1, 130-148.

Burdick, H.O. and Pincus, G. 1935. The effect of estrin injections upon the developing ova of mice and rabbits. Am. J. Physiol., 111, 201-208.

Chang, M.C. 1950. Development and fate of transferred rabbit ova or blastocyst in relation to the ovulation time of recipients. J. exp. Zool., 114, 197-216.

Chang, M.C. 1976. Estrogen, progesterone and egg transport - overview and identification of problems. In "Ovum Transport and Fertility Regulations" (WHO Symposium, San Antonio, Texas Scriptor, Copenhagen). pp. 473-484.

Day, B.N. and Polge, C. 1968. Effects of progesterone on fertilization and egg transport in the pig. J. Reprod. Fert., 17, 227-230.

Dey, S.K., Johnson, D.C. and Santos, J.G. 1979a. Is histamine production by the blastocyst required for implantation in the rabbit Biol. Reprod., 21, 1169-1173.

Dey, S.K., Villanueva, C. and Abdou, N.I. 1979b. Histamine receptors on rabbit blastocyst and endometrial cell membranes. Nature (Lond.), 278, 648-649.

Dubois, W., Brackett, K.H., Bar-Ilan, D. and Fields, P.A. 1983. Peptide synthesis by the bovine corpus luteum. Biol. Reprod., 28, Suppl. 1, 77.

Edey, T.N. 1979. Embryo mortality. In "Sheep Breeding" (Ed. G.J. Tomes, D.E. Robertson and R.J. Lightfoot). 2nd edition. (Butterworths, London). pp. 315-325.

Ellinwood, W.E. 1978. Maternal recognition of pregnancy in the ewe and the rabbit (PhD Thesis, Colorado State University).

Ellinwood, W.E., Nett, T.M. and Niswender, G.D. 1979. Maintenance of the corpus luteum of early pregnancy in the ewe. I. Luteotropic properties of embryonic homogenates. Biol. Reprod., 21, 281-288.

Feigelson, M., and Kay, E. 1972. Protein patterns of rabbit oviducal fluid. Biol. Reprod., 6, 244-252.

Fields, P.A., Eldridge, R.K., Fuchs, A.-R., Roberts, R.F. and Fields, M.J. 1983. Human placental and bovine corpora luteal oxytocin. Endocrinology, 112, 1544-1546.

Findlay, J.K. 1983. The endocrinology of the preimplantation period. Current Topics exp. Endocr. 4, 35-67.

Findlay, J.K., Ackland, N., Burton, R.D., Davis, A.J., Maule Walker, F.M., Walters, D.E. and Heap, R.B. 1981. Protein, prostaglandin and steroid synthesis in caruncular and intercaruncular endometrium of sheep before implantation. J. Reprod. Fert., 62, 361-377.

Fleet, I.R. and Heap, R.B. 1982. Uterine blood flow, myometrial activity and their response to adenosine during the peri-implantatiuon period in sheep. J. Reprod. Fert., 65, 195-205.

Flint, A.P.F. and Sheldrick, E.L. 1982. Ovarian secretion of oxytocin is stimulated by prostaglandin. Nature (Londn.), 297, 587-588.

Flint, A.P.F. and Sheldrick, E.L. 1983. Evidence for a systemic role for ovarian oxytocin in luteal regression in sheep. J. Reprod. Fert., 67, 215-225.

Flint, A.P.F., Burton, R.D., Gadsby, J.E., Saunders, P.T.K. and Heap, R.B. 1979. Blastocyst oestrogen synthesis and the maternal recognition of pregnancy. In "Maternal Recognition of Pregnancy", Ciba Foundation Symposium 64 (new series) (Ed. J. Whelan). (Excerpta Medica, Amsterdam). pp. 209-228.

Flint, A.P.F., Saunders, P.T.K. and Ziecik, A.J. 1982. Blastocyst-endometrium interactions and their significance in embryonic mortality. In "Control of Pig Production" (Ed. D.J.A. Cole and G.R. Foxcroft). (Butterworths, London). pp. 253-275.

Ford, S.P., Chenault, J.R. and Echternkamp, S.E. 1979. Uterine blood flow of cows during the oestrus cycle and early pregnancy: effect of the conceptus on the uterine blood supply. J. Reprod. Fert., 56, 53-62.

Ford, S.P. and Christenson, R.K. 1979. Blood flow to uteri of sows during the estrous cycle and early pregnancy: local effect of the conceptus on the uterine blood supply. Biol. Reprod., 21, 617-624.

Ford, S.P., Christenson, R.K., Ford, J.J. 1982. Uterine blood flow and uterine arterial, venous and luminal concentrations of oestrogens on Days 11, 13 and 15 after oestrus in pregnant and non-pregnant sows. J. Reprod. Fert., 64, 185-190.

Gadsby, J.E., Heap, R.B. and Burton, R.D. 1980. Oestrogen production by blastocyst and early embryonic tissue of various species. J. Reprod. Fert., 60, 409-417.

Gardner, M.L., First, N.L. and Casida, L.E. 1963. Effect of exogenous estrogens on corpus luteum maintenance in gilts. J. Anim. Sci., 22, 132-134.

Garverick, H.A., Polge, C. and Flint, A.P.F. 1982. Oestradiol administration raises luteal LH receptor levels in intact and hysterectomized pigs. J. Reprod. Fert., 66, 371-377.

Geisert, R.D., Brookbank, J.W., Roberts, R.M. and Bazer, R.W. 1982. Establishment of pregnancy in the pig. II. Cellular remodelling of the porcine blastocyst during elongation on day 12 of pregnancy. Biol. Reprod., 27, 941-955.

Godkin, J.D., Cote, C. and Duby, R.T. 1978. Embryonic stimulation of ovine and bovine corpora lutea. J. Reprod. Fert., 54, 375-378.

Godkin, J.D., Bazer, W.W., Thatcher, W.E. and Roberts, R.M. 1984. Proteins released by cultured day 15-16 conceptuses prolong luteal maintenance when introduced into the uterine lumen of cyclic ewes. J. Reprod. Fert., 71, 57-64.

Godkin, J.D., Bazer, F.W., Moffatt, J., Sessions, F. and Roberts, R.M. 1982. Purification and properties of a major low molecular weight protein released by the trophoblast of sheep blastocysts at Day 13-21. J. Reprod. Fert., 65, 141-150.

Greenwald, G.S. 1958. Endocrine regulation of the secretion of mucin in the tubal epithelium of the rabbit. Anat. Rec., 130, 477-495.

Greenwald, G.S. 1967. Species differences in egg transport in response to exogenous estrogen. Anat. Rec., 157, 163-172.

Greenwald, G.S. 1976. Effects of estrogen and progesterone on egg transport: Summary and conclusions. In "Ovum Transport and Fertility Regulation" (WHO Symposium, San Antonio). pp. 539-543.

Guillomot, M., Fléchon, J.E. and Wintenberger-Torrès, S. 1981. Conceptus attachment in the ewe: an ultrastructure study. Placenta, 2, 169-182.

Guillomot, M., Fléchon, J.E. and Wintenberger-Torrès, S. 1982a. Cyto-chemical studies of the uterus and trophoblastic surface coats during blastocyst attachment in the ewe. J. Reprod. Fert., 65, 1-8.

Guillomot, M. and Guay, P. 1982b. Ultrastructural features of the cell surfaces of uterine and trophoblastic epithelium during embryonic attachment in the cow. Anat. Rec., 204, 315-322.

Gustavsson, I. 1984. Chromosome evaluation and fertility. 10th Int. Cong. Anim. Reprod. Art. Insem. vol. VI (University of Illinois at Urbana-Champaign, Illinois). pp. 1-8.

Hamilton, D.W., Allen, W.R. and Moor, R.M. 1973. Origin of endometrial cups. III. Light and electron microscopic study of fully developed equine endometrial cups. Anat. Rec., 177, 503-518.

Hamner, C.E. 1973. Oviducal fluid - composition and physiology. In "Handbook of Physiology" Section 7. Endocrinology 2 part II (Ed. R.O. Greep). (American Physiological Society, Washington). pp. 141-151.

Hamner, C.E. and Fox, S.B. 1969. Biochemistry of oviductal secretions. In "The Mammalian Oviduct" (Ed. E.S.E. Hafez and R.J. Blandau). (University of Chicago Press, Chicago). pp. 333-355.

Hanly, S. 1961. Prenatal mortality in farm animals. J. Reprod. Fert., 2, 182-194.

Hansel, N. and Fortune, J. 1978. The applications of ovulation control. In "Control of Ovulation" (Ed. D.B. Crighton, N.B. Haynes, G.R. Foxcroft and G.E. Lamming). (Butterworths, London). pp. 237-263.

Harrison, F.A. 1979. Luteolysin in the pig. J. Physiol. Lond., 290, 36P.

Harrison, F.A., Heap, R.B. and Poyser, N.L. 1976. Production, chemical composition and prostaglandin $F_{2\alpha}$ content of uterine fluid in pregnancy sheep. J. Reprod. Fert., 48, 61-67.

Heap, R.B., Flint, A.P.F. and Gadsby, J.E. 1981. Embryonic signals and maternal recognition. In "Cellular and Molecular Aspects of Implantation" (Ed. S.R. Glasser and D.W. Bullock). (Plenum Press, New York). pp. 311-326.

Heap, R.B., Hamon, M. and Allen, W.R. 1982. Studies on oestrogen synthesis by the preimplantation equine conceptus. J. Reprod. Fert., Suppl. 32, 343-352.

Heap, R.B., Staples, L.D., Flint, A.P.F., Maule Walker, F.M. and Allen, W.R. 1983. Synthetic capabilities of the preimplantation conceptus and maternal responses to pregnancy. In "Fertilization of the Human Egg In Vitro - Biological Basis and Clinical Implications" (Ed. H.M. Beier and H. Lindner). (Springer Verlag, Berlin). pp. 387-411.

Horton, E.W. and Poyser, N.L. 1976. Uterine luteolytic hormone : a physiological role for prostaglandin $F_{2\alpha}$. Physiological Reviews, 56, 595-651.

Humphrey, K.W. 1976. Influence of estrogen and progesterone on oviductal function. In "Ovum Transport and Fertility Regulation" (WHO Symposium, San Antonio, Texas, Scriptor - Copenhagen). pp. 485-505.

Hunter, R.H.F. 1977. Physiological factors influencing ovulation

fertilisation, early embryonic development and establishment of
pregnancy in the pig. Brit. vet. J., _133_, 461-470.

Hunter, R.H.F. 1980. "Physiology and Technology of Reproduction in Female
Domestic Animals". (Academic Press, London). pp. 145-183.

Johnson, M.H. 1979. Intrinsic and extrinsic factors in preimplantion
development. J. Reprod. Fert., _55_, 255-265.

Kennedy, T.G. 1980. Prostaglandins and the endometrial vascular
permeability changes preceding blastocyst inmplantation and
decidualization. In " Blastocyst-Endometrium Relationships" (Ed. F.
Leroy, C.A. Finn, A. Psychoyos). (Karger-Basel). pp. 234-243.

Kidder, H.E., Casida, L.E. and Grummer, R.H. 1955. Some effects of
estrogen injections on the estrual cycle of gilts. J. Anim. Sci., _14_,
470-474.

King, G.J., Atkinson, B.A. and Robertson, H.A. 1980. Development of the
bovine placentome from days 20 to 29 of gestation. J. Reprod. Fert.,
_59_, 95-100.

King, Atkinson, B.A. and Robertson, H.A. 1981. Development of the
intercarruncular areas during early gestation and the establishment of
the bovine placenta. J. Reprod. Fert., _61_, 469-474.

Kirby, D.R.S. 1962. The influence of the uterine environment on the
development of mouse eggs. J. Embryol. exp. Morph., _10_, 496-506.

Knickerbocker, J.J., Thatcher, W.W., Bazer, F.W., Drost, M., Barron, D.H.,
Fincher, K.B. and Roberts, R.M. 1984. Proteins secreted by cultured
day 17 bovine conceptuses extend luteal function in cattle. 10th Int.
Cong. Anim. Reprod. Art. Insem., vol. _2._ (University of Illinois at
Urbana-Champaign). p. 88.

Koch, E. 1985. Early pregnancy factor: its significance as an indicator of
fertilization and embryonic viability. (In this volume).

Lambert, J.C., Hamner, C.E. and Gemmill, C.L. 1973. An ultrastructural
study of secretion in the rabbit oviduct under different hormonal
influence. Fed. Proc., _32_, 213A.

Lauritsen, J.G. 1982. The cytogenetics of spontaneous abortion. Res. Rev.,
_14_, 3-4.

Lee, S.Y., Mossman, H.W., Mossman, A.S. and Del Pino, G. 1977. Evidence of
a specific nidation site in ruminants. Am. J. Anat., _150_, 631-640.

Leese, H.J. and Aldridge, S. 1979. The movement of pyruvate, lactate
dehydrogenase into rabbit oviductal fluid. J. Reprod. Fert., _56_,
619-622.

Leiser, R. 1975. Kontaktanfuahuse zuischen Trophoblast und Uterusepithel
wahrend der fruhen Implantation beim Rind. Anat. Hist. Embryol., _4_,
63-86.

Maas, D.H.A., Storey, B.T. and Mastroianni, L. Jr. 1976. Oxygen tension in
the oviduct of the rhesus monkey. Fert. Steril., _27_, 1312-1317.

Martal, J., Djiane, J. and Dubois, M.P. 1977. Immunofluorescent
localisation of ovine placental lactogen. Cell and Tissue Research,
_184_, 427-433.

Martal, J., Lacroix, M.-C., Loudes, C., Saunier, M. and Winterberger-
Torrès, S. 1979. Trophoblastin, an anti-luteolytic protein present in
early pregnancy in sheep. J. Reprod. Fert., _56_, 63-71.

Masters, R.A., Roberts, R.M., Lewis, G.S., Thatcher, W.W., Bazer, F.W. and
Godkin, J.D. 1982. High molecular weight glycoproteins released by
expanding, pre-attachment sheep, pig and cow blastocysts in culture.
J. Reprod. Fert., _66_, 571-583.

Mastroianni, L. Jr. and Go, K.J. 1979. Tubal secretions. In "The Biology
of the Fluids of the Female Genital Tract" (Ed. F.K. Beller and G.F.B.

Schumacher). (Elsevier, North Holland, New York). pp. 335-344.

Mastroianni, L. Jr. and Jones, R. 1965. Oxygen tension within the rabbit fallopian tube. J. Reprod. Fert., 9, 99-102.

Mastroianni, L. Jr., Urzua, M. and Stambaugh, R. 1970. Protein patterns in monkey oviductal fluid before and after ovulation. Fert. Steril., 21, 817-820.

Mastroianni, L. Jr., Urzua, M. and Stambaugh, R. 1973. The internal environmental fluids of the oviduct. In "The Regulation of Mammalian Reproduction" (Ed. S.J. Segal, R. Crozier, P.A. Corfman and R.G. Condliffe). (Charles C. Thomas, Springfield). pp. 376-381.

Maule Walker, F.M. 1983. Lactation and fertility in goats after tne induction of parturition with an analogue of prostaglandin $F_{2\alpha}$, cloprostenol. Res. vet. Sci., 34, 280-286.

McCracken, J.A., Carlson, J.C., Glew, M.E., Goding, J.R., Baird, D.T., Green, K. and Samuelsson, B. 1972. Prostaglandin F_2 identified as a luteolytic hormone in the sheep. Nature, New Biol., 238, 129-134.

McCracken, J.A., Schramm, W. and Okulicz, W.C. 1984. Hormone receptor control of pulsatile secretion of $PGF_{2\alpha}$ from the ovine uterus during luteolysis and its abrogation in early pregnancy. Anim. Reprod. Sci., 7, 31-56.

McCracken, J.A., Schramm, W., Barcikowski, B. and Wilson, L. 1981. The identification of prostaglandin $F_{2\alpha}$ as a uterine luteolytic hormone in the sheep and the endocrine control of its synthesis. Acta vet. scand., Suppl. 77, 71-88.

McDonald, M.F. and Bellvé, A.R. 1969. Influence of oestrogen and progesterone on flow of fluid from the Fallopian tube in the ovariectomized ewe. J. Reprod. Fert., 20, 51-61.

Moeljono, M.P.E., Bazer, F.W. and Thatcher, W.W. 1976. A study of prostaglandin $F_{2\alpha}$ as the luteolysin in swine: I. Effect of prostaglandin $F_{2\alpha}$ in hysterectomized gilts. Prostaglandins, 11, 737-743.

Moeljono, M.P.E., Thatcher, W.W., Bazer, F.W., Frank, M., Owens, M.J. and Wilcox, C.J. 1977. A study of prostaglandin $F_{2\alpha}$ as the luteolysin in swine: II. Characterization and comparison of prostaglandin F_2, oestrogen and progestin concentrations in utero-ovarian vein plasma of non-pregnant and pregnant gilts. Prostaglandins, 14, 543-555.

Moor, R.M. 1968. Effect of embryo on corpus luteum function. J. Anim. Sci., 27, Suppl. 1, 97-118.

Morton, H., Hegh, V. and Clunie, G.J.A. 1976. Studies of the rosette inhibition test in pregnant mice: evidence of immunosuppression. Proc. R. Soc. Lond. B., 193, 413-419.

Morton, H., Morton, D.J. and Ellendorff, F. 1983. The appearance and characteristics of early pregnancy factor in the pig. J. Reprod. Fert., 69, 437-466.

Morton, H., Rolfe, B., Clunie, G.J.A., Anderson, M.J. and Morrison, J. 1977. An early pregnancy factor detected in human serum by the rosette inhibition test. Lancet, i, 394-397.

Murray, R.A., Bazer, F.W. Rundell, J.W., Vincent, C.K., Wallace, H.D. and Warnick, A.C. 1971. Developmental failure of swine embryos restricted to the oviducal environment. J. Reprod. Fert., 24, 445-448.

Nancarrow, C.D., Buckmaster, J., Chamley, W., Cox, R.I., Cumming, I.A., Cummins, L., Drinan, J.P., Findlay, J.K., Goding, J.R., Restall, B.J., Schneider, W. and Thorburn, G.D. 1973. Hormonal changes around oestrus in the cow. J. Reprod. Fert., 32, 320-321.

Nieder, G.L. and Corder, C.L. 1982. Qualitative histochemical measurement of pyruvate and lactate in the mouse oviduct during the estrus cycle. J. Histochem. Cytochem., 30, 1051-1058.

Noske, I.G. and Daniel, J.C. Jr. 1974. Changes in uterine and oviducal fluid proteins during early pregnancy in the golden hamster. J. Reprod. Fert., 38, 173-176.

Oliphant, G. and Ross, P.R. 1982. Demonstration of production and isolation of three sulfated glycoproteins from the rabbit oviduct. Biol. Reprod., 26, 537-544.

Oliphant, G. Bowling, A., Eng., L.A., Keen, S. and Randall, P.A. 1978. The permeability of the rabbit oviduct to proteins present in the serum. Biol. Reprod., 18, 516-520.

Oliphant, G., Reynolds, A.B., Smith, P.F., Ross, P.R. and Marta, J.S. 1984. Immunocytochemical localization and determination of hormone-induced synthesis of the sulfated oviductal glycorproteins. Biol. Reprod., 31, 165-174.

Ottobre, J.S., Vincent, D.L., Silvia, W.J. and Inskeep, E.K. 1984. Aspects of regulation of uterine secretion of prostaglandins during the oestrous cycle and early pregnancy. Anim. Reprod. Sci., 7, 75-100.

Perkins, J.L. 1974. Fluid flow of the oviduct. In "The Oviduct and its Functions" (Ed. A.D. Johnson and C.W. Foley). (Academic Press, New York). pp. 119-132.

Perry, J.S. and Rowlands, I.W. 1962. Early pregnancy in the pig. J. Reprod. Fert., 4, 175-188.

Perry, J.S. 1960. The incidence of embryonic mortality as a characteristic of the individual sow. J. Reprod. Fert., 1, 71-83.

Perry, J.S., Heap, R.B. and Amoroso, E.C. 1973. Steroid hormone production by pig blastocysts. Nature (Lond.), 245, 45-47.

Pharriss, B.B. and Wyngarden, J.J. 1969. The effect of prostaglandin $F_{2\alpha}$ on the progesterone content of ovaries from pseudopregnant rats. Proc. Soc. exp. Biol. Med., 130, 92-94.

Pope, C.E. and Day, B.N. 1972. Development of pig embryos following restriction to the ampullar portion of the oviduct. J. Reprod. Fert., 31, 135-138.

Psychoyos, A. 1966. Etude des relations de l'oeuf et de l'endomètre au cours du retard de la nidation ou des premières phases du processus de nidation chez la rutte. C.R. hebd. Seanc. Acad. Sci., Paris Ser D., 263, 1755-1758.

Restall, B.J. 1965. The fallopian tube of the sheep. II. The influence of progesterone and oestrogen on the secretory activities of the fallopian tube. Aust. J. biol. Sci., 19, 187-197.

Richardson, L.L. and Oliphant, G. 1981. Steroid concentrations in rabbit oviducal fluid during oestrus and pseudopregnancy. J. Reprod. Fert., 62, 427-431.

Ricketts, A.P., Scott, D.W. and Bullock, D.W. 1984. Radioiodinated surface proteins of separated cell types from rabbit endometrium in relation to the time of implantation. Cell and Tissue Res., 236, 421-429.

Rider, V., MacRae, A., Heap, R.B. and Feinstein, A. 1985. Passive immunization against progesterone inhibits endometrial sensitization in pseudopregnant mice and has anti-fertility effects in pregnant mice which are reversible by steroid treatment. J. Endocr. (in press).

Roberts, G.P., Parker, J.M. and Symonds, H.W. 1975. Proteins in the luminal fluid from the bovine oviduct. J. Reprod. Fert., 45, 301-313.

Robertson, H.A. and King, G.J. 1974. Plasma concentrations of progesterone, oestrone, oestradiol-17β and of oestrone sulphate in the pig at implantation, during pregnancy and at parturition. J. Reprod.

71

Fert., <u>40</u>, 133-141.
Roblero, L.S. and Garavayno, A.C. 1979. Effect of oestradiol-17β and progesterone on oviductal transport and early development of mouse embryos. J. Reprod. Fert., <u>57</u>, 91-95.
Roblero, L., Biggers, J.D. and Lechene, C.P. 1976. Electron probe analysis of the elemental microenvironment of oviducal mouse embryos. J. Reprod. Fert., <u>46</u>, 431-434.
Roche, J.F., Borland, M.P. and McGrady, T.A. 1981. Reproductive wastage following artificial insemination of heifers. Vet. Rec., <u>109</u>, 401-404.
Rowson, L.E.A. and Moor, R.M. 1967. The influence of embryonic tissue homogenate infused into the uterus on the life-span of the corpus luteum in sheep. J. Reprod. Fert., <u>13</u>, 511-516.
Ryan, J.P. and Moore, N.W. 1983. Transport and development of embryos transferred to the oviduct of entire and ovariectomized ewes. Proc. 15th Ann. Conf. of Aust. Soc. Reprod. Biol., Canberra, p. 93.
Saba, N. and Hattersley, J.P. 1981. Direct estimation of oestrone sulphate in sow serum for a rapid pregnancy diagnosis test. J. Reprod. Fert., <u>62</u>, 87-92.
Sacco, A.G. and Shivers, C.A. 1973. Comparison of antigens in the ovary, oviduct and uterus of the rabit and other mammalian species. J. Reprod. Fert., <u>32</u>, 421-427.
Samuel, C.A. and Perry, J.S. 1972. The ultrastructure of pig trophoblast transplanted to an ectopic site in the uterine wall. J. Anat., <u>113</u>, 139-149.
Schlafke, S. and Enders, A.C. 1975. Cellular basis of interaction between trophoblast and uterus at implantation. Biol. Reprod., <u>12</u>, 41-65.
Schramm, W., Bovaird, L., Glew, M.E., Schramm, G. and McCracken, J.A. 1983. Corpus luteum regression induced by ultra-low pulses of prostaglandin $F_{2\alpha}$. Prostaglandins, <u>26</u>, 347-364.
Schramm, D., Prokopp, S. and Barth, D. 1983. The effect of active and passive immunization against oxytocin on ovarian cyclicity in ewes. Acta endocr. Copenhagen, <u>103</u>, 337-344.
Shapiro, S.S., Jentsch, J.R. and Yard, A.S. 1971. Protein composition of rabbit oviduct fluid. J. Reprod. Fert., <u>24</u>, 403-408.
Sheldrick, E.L. and Flint, A.P.F. 1983a. Luteal concentrations of oxytocin decline during early pregnancy in the ewe. J. Reprod. Fert., <u>68</u>, 477-480.
Sheldrick, E.L. and Flint, A.P.F. 1983b. Regression of the corpora lutea in sheep in response to cloprostenol is not affected by loss of luteal oxytocin after hysterectomy. J. Reprod. Fert., <u>68</u>, 155-160.
Sheldrick, E.L., Mitchell, M.D. and Flint, A.P.F. 1980. Delayed luteal regression in ewes immunized against oxytocin. J. Reprod. Fert., <u>59</u>, 37-42.
Short, R.V. 1969. Implantation and the maternal recognition of pregnancy. In "Foetal Autonomy" Ciba Foundation Symposium, (Ed. G.E.W. Wolstenholme and Maeve O'Connor). (Churchill, London). pp. 2-26.
Short, R.V. 1979. When a conception fails to become a pregnancy. In "Maternal Recognition of Pregnancy", Ciba Foundation Symposium 64 (new series) (Ed. J. Whelan). (Excerpta Medica, Amsterdam). pp. 377-394.
Silvia, W.J., Fitz, T.A., Mayan, M.H. and Niswender, G.D. 1984. Cellular and molecular mechanisms involved in luteolysis and maternal recognition of pregnancy in the ewe. J. Anim. Reprod. Sci., <u>7</u>, 57-74.
Singh, M.M., Wadhwa, V., Sethi, N. and Kamboj, V.P. 1983. Viability and development of 'tube locked' mouse embryos. J. Reprod. Fert., <u>68</u>,

165-170.

Sreenan, J.M. and Diskin, M.G. 1983. Early embryonic mortality in the cow: its relationship with progesterone concentration. Vet. Rec., 112, 517-521.

Stone, S.L., Huckle, W.R. and Oliphant, G. 1980. Identification and hormonal control of reproductive-tract-specific antigens present in rabbit oviductal fluid. Gamete Res., 3, 169-177.

Surani, M.A.H. and Fishel, S.B. 1980. Blastocyst uterine interactions at implantation. Prog. Reprod. Biol., 7, 14-17.

Sutton, R., Wallace, A.L. and Nancarrow, C.D. 1983. EGP: a glycoprotein present in sheep oviduct fluid at oestrus. Proc. 15th Ann. Conf. Aust. Soc. Reprod. Biol., Canberra, p. 103.

Thatcher, W.W., Wolfenson, D., Curl, J.S., Rico, L.E., Knickerbocker, J.J., Bazer, F.W. and Drost, M. 1984. Prostaglandin dynamics associated with development of the bovine conceptus. J. Anim. Reprod. Sci. 7, 149-176.

Tyndale-Biscoe, C.H. 1979. Hormonal control of embryonic diapause and reactivation in the tammar wallaby. In "Maternal Recognition of Pregnancy" Ciba Foundation Symposium 64 (new series), (Ed. J. Whelan). (Excerpta Medica, Amsterdam). pp. 173-190.

van Niekerk, C.H. and Gernecke, W.H. 1966. Persistence and parthenogenetic cleavage of tubal ova in the mare. Onderstepoort J. vet. Res., 33, 195-232.

Wang, M.-Y., Rider V., Heap, R.B. and Feinstein, A. 1984. Action of anti-progesterone monoclonal antibody in blocking pregnancy after post coital administration in mice. J. Endocr., 101, 95-100.

Wathes, D.C. and Swann, A.B. 1982. Is oxytocin an ovarian hormone? Nature (Lond.), 297, 225-227.

Wathes, D.C., Swann, R.W., Birkett, S.D., Porter, D.G. and Pickering, B.T. 1983. Characterization of oxytocin, vasopressin and neurophysin from the bovine corpus luteum. Endocrinology, 113, 693-698.

Wathes, D.C. and Wooding, F.B.P. 1981. An electron microscopic study of implantation in the cow. Am. J. Anat., 159, 285-306.

Watkins, W.B., Moor, L.G., Flint, A.P.F. and Sheldrick, E.L. 1983. Secretion of neurophysins by the ovary in sheep. Peptides, 5, 61-64.

Watkins, W.B. and Reddy, S. 1980. Ovine placental lactogen in the cotyledonary and intercotyledonary placenta of the ewe. J. Reprod. Fert., 58, 411-414.

Watson, J. and Maule Walker, F.M. 1977. Effect of prostaglandin F-2α and uterine extracts on progesterone secretion in vitro by superfused pig corpora lutea. J. Reprod. Fert., 51, 393-398.

Weitlauf, H.M. 1971. Protein synthesis by blastocysts in the uteri and oviducts of intact and hypophysectomized mice. J. exp. Zool., 176, 35-40.

Wettemann, R.P., Bazer, F.W., Thatcher, W.W. and Hoagland, T.A. 1984. Environmental influences on embryonic mortality. 10th Int. Cong. Anim. Reprod. Art. Insem., vol. 4, (University of Illinois at Urbana-Champaign, Illinois) XIII-26 - XIII-32.

Whittingham, D.G. 1968. Development of zygotes in cultured mouse oviducts. II. The influence of the estrous cycle and ovarian hormones upon the development of the zygote. J. exp. Zool., 169, 399-406.

Wimsatt, W.A. 1975. Some comparative aspects of implantation. Biol. Reprod., 12, 1-40.

Wintenberger-Torrès, S. 1956. Les rapports entre l'oeuf en segmentation et le tractus maternel chez la brebis. Third Int. Cong. Anim. Reprod.

(Camb.), 1, 62-64.

Wintenberger-Torrès, S. and Fléchon, J.E. 1974. Ultrastructure evolution of the trophoblast cells of the preimplantation sheep blastocyst from day 8 to 18. J. Anat., 118, 143-153.

Wooding, F.B.P. 1981. Localisation of ovine placental lactogen in sheep placentomes by electron microscope immunocytochemistry. J. Reprod. Fert., 62, 15-19.

Wooding, F.B.P. 1982. The role of the binucleate cell in ruminant placentae structure. J. Reprod. Fert., Suppl. 31, 31-39.

Wooding, F.B.P. 1983. Frequency and localisation of binucleate cells in the placentomes of ruminants. Placenta 4, 527-540.

Wooding, F.B.P. 1984. Role of binucleate cells in fetomaternal cell fusion at implantation in the sheep. Am. J. Anat., 170, 233-250.

Wooding, F.B.P., Staples, L.D. and Peacock, M.A. 1982. Structure of trophoblast papillae on the sheep conceptus at implantation. J. Anat., 134, 507-516.

Wooding, F.B.P. and Wathes, D.C. 1980. Binucleate cell migration in the bovine placentome. J. Reprod. Fert., 59, 425-430.

Wright, L.J., Feinstein, A., Heap, R.B., Saunders, J.C., Bennett, R.C. and Wang, M.-Y. 1982. Anti-progesterone monoclonal antibody locks implantation in mice. Nature (Lond.), 295, 415-417.

Wu, C., Mastroianni, L. Jr. and Mikhail, G. 1977. Steroid hormones in monkey oviduct fluid. Fert. Steril., 28, 1250-1256.

EARLY PREGNANCY FACTOR : ITS SIGNIFICANCE AS AN INDICATOR OF FERTILIZATION AND EMBRYONIC VIABILITY

E. Koch

Institut für Tierzucht und Tierverhalten, FAL,
Mariensee, 3057 Neustadt 1, F.R. Germany and
AFRC Institute of Animal Physiology,
Babraham, Cambridge CB2 4AT, U.K.

ABSTRACT

Early pregnancy factor (EPF) is a pregnancy-associated protein complex and has been detected in sera of mice, humans, sheep, cattle and pigs. EPF is demonstrated by the rosette inhibition test, in which it possesses a synergistic effect with anti-lymphocyte serum in preventing rosette formation between T-lymphocytes and heterologous erythrocytes. Its biological function is thought to be one of immunoregulation during pregnancy. The initial stimulation and sustained production of EPF is claimed to be specific for fertilization and the continuing presence of a viable conceptus, thus raising the possibility of continuous monitoring during the course of early pregnancy. This review briefly summarizes a substantial part of the available literature on EPF as well as my own unpublished work.

INTRODUCTION

The occurrence and significance of embryonic mortality in farm animals are well-known (for reviews in cattle, sheep and pig see Ayalon, 1978; Edey, 1979, Flint et al., 1982). However, so far no clinical method is available which permits the detection or even the temporal determination of embryonic death in a conscious animal. As a high percentage of conceptuses is lost before pregnancy diagnosis is possible or before any fluctuation of the oestrous cycle is obvious, the majority of miscarriages remains completely unrecognized. The assessment of this problem is only possible by slaughter of representative numbers of animals at different stages of pregnancy. Thus, a non-surgical technique which allows both detection of fertilization and continuous monitoring of the course of pregnancy would be most valuable for the study of embryonic mortality.

The earliest known parameter claimed to be specific for fertilization and the continuing presence of a viable conceptus is a serum constituent which has originally been detected in mice (Morton et al., 1976, Morton et al., 1982a). This substance, termed early pregnancy factor (EPF), has also been described in humans (Morton et al., 1977; Smart et al., 1982), sheep (Morton et al., 1979a,b), cattle (Nancarrow et al., 1981) and pigs (Paisley et al., 1982; Morton et al., 1983). The reported extraordinary properties of EPF - early appearance after mating or insemination and rapid disappearance following induced death or removal of embryos - suggest it to be a most useful tool for the investigation of early pregnancy as well as its failure (Rolfe, 1982).

The present paper reviews the work on EPF with special relevance to the study of early pregnancy and embryonic mortality in domestic animals.

DETECTION OF EPF-ACTIVITY

The detection of EPF is entirely dependent on the use of the rosette inhibition test. This test is based on the ability of anti-lymphocyte serum (ALS) to inhibit the formation of spontaneous rosettes between T-lymphocytes and heterologous red blood cells (Bach and Antoine, 1968) (Fig. 1). Employing the rosette inhibition test for the evaluation of ALS Morton et al. (1974) discovered that rosettes formed with lymphocytes from pregnant mice during the first 2 weeks of pregnancy could be inhibited

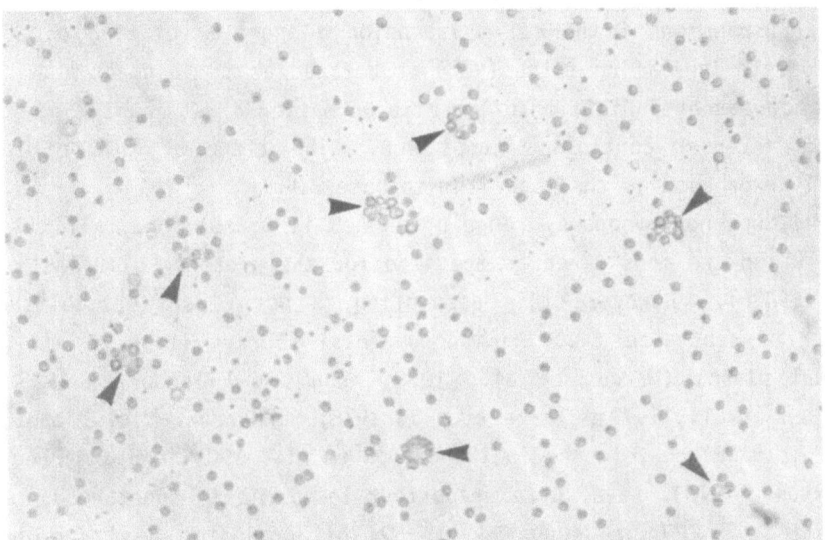

Fig. 1   T-lymphocytes of many species are characterized by their ability to form cell clusters with heterologous erythrocytes, called spontaneous or E-rosettes. The inhibition of rosette formation is the means for the detection of EPF. The microphotograph (x320) shows the formation of E-rosettes between pig lymphocytes and sheep red blood cells (arrows).

with significantly lower dilutions of a standard-ALS as compared to lymphocytes of non-pregnant animals. Since this reaction occurred within 4-6 h after mating and could also be induced with normal lymphocytes after preincubation in pregnancy serum (Morton et al., 1976), they concluded that a pregnancy-dependent serum constituent exists, which was later termed early pregnancy factor (EPF) (Morton et al., 1977). As EPF is exclusively characterized by the increased rosette inhibition titre of a particular ALS, this effect will be called EPF-activity in this paper.

A schematic representation of the rosette inhibition test as adopted for the detection of EPF-activity in pig serum is shown in Fig. 2. The test is performed as follows: pig lymphocytes are incubated in test serum and known EPF-positive and EPF-negative control samples for 30 min at 37°C. The cells (100 μl; 15,000 cells/μl) are subsequently washed twice and added to a series of 10 different ALS-dilutions (250 μl, 4-fold diluted from 1/1000 to $1/250 \times 10^6$ in Hanks balanced salt solution, HBSS) and guinea-pig serum (50 μl, diluted 1:5). This mixture is incubated for 90 min. Formation of rosettes is achieved by the addition of 100 μl sheep red-blood-cell suspension (SRBC, 10^5 cells/μl), centrifugation for 5 min and resuspension of the cell suspension by gentle rotation on a wheel. The rosette inhibition titre (RIT) is defined as the highest ALS-dilution which causes an inhibition of rosette formation of at least 25%, compared to two internal controls without ALS. The titre of this dilution is usually expressed as the logarithm to base 2.

If the non-pregnancy range of an ALS is established, all RITs above this value are considered as specific for the presence of EPF (Morton et al., 1982b). However, this assumption is not clearly established. A number of substances, e.g. immunosuppressive drugs (Bach et al., 1969), seminal plasma (Marcus et al., 1979; Koch and Ellendorff, 1985b), hCG (Morton et al., 1977; Clarke et al., 1978; Nancarrow et al., 1981; Rolfe et al., 1983a), α-fetoprotein (Nancarrow et al., 1981) and α-macroglobulin (Stimson, 1976) have been reported to inhibit spontaneous rosette formation. EPF is supposed to inhibit rosette formation only in cooperation with ALS and this has been shown to be true for serum of pregnant pigs (Koch and Ellendorff, 1985a). Most of the above mentioned substances, in contrast, possess inherent rosette-inhibiting activity. Thus, if serum or other body fluids and tissues contain two or more constituents which affect rosette formation, the importance of each

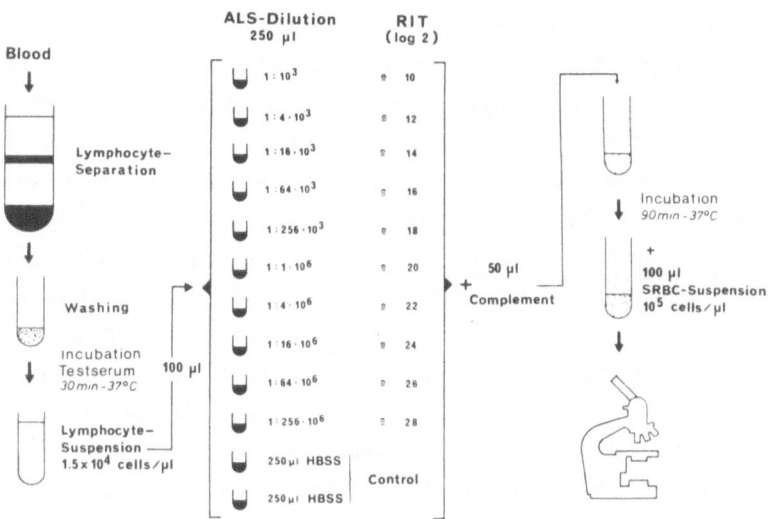

Fig. 2 Schematic representation of the rosette inhibition test as adapted for the detection of EPF-activity in swine serum (after Smart et al., 1982).

compound will require further study. With respect to this problem Rolfe et al. (1983a) demonstrated that EPF is a contaminant of some commercial hCG preparations and suggest that EPF, not hCG, is responsible for the immunosuppressive properties of these preparations. Seminal plasma, on the other hand, has been shown to contain at least two different components: a dialyzable and heat-labile substance with inherent rosette inhibiting activity, and a non-dialyzable and heat-stable (56°C) constituent, which behaves like EPF and prevents rosette formation only in cooperation with ALS (Koch and Ellendorff, 1985b).

Validation of the rosette inhibition test

Although several independent groups confirmed the observations of Morton and her co-workers in different species (Nancarrow et al., 1981; Grewal et al., 1981; Paisley et al., 1982; Smart et al., 1982) there was a constant awareness of the inadequacies of the test system (Paisley et al., 1982; Smart et al., 1982; Koch et al., 1983). As some authors were entirely unsuccessful in repeating the results in humans (Thomson et al., 1980; Cooper and Aitken, 1981) and a complete validation of the rosette inhibition test is still lacking, the assay system as well as EPF have remained subjects of controversial discussion (Whyte and Heap, 1983).

Failures of the rosette inhibition test are usually attributed to alterations from the original method (Morton et al., 1982b). It is important to mention that rosettes active in this test are only those which form immediately after addition of erythrocytes without any enhancement. In the pig, for example, the percentage of these cells is about 10% (Koch and Ellendorff, 1985a). The number of rosette-forming cells can gradually be increased up to some 60% by incubation of cells in the presence of dextran, fetal calf serum, Ficoll 400 or polyethylene glycol, by incubation at 4°C overnight as well as by treatment of red blood cells with papain or 2-amino ethyl isothiouronium bromide (AET) (Binns, 1982). However, these enhanced rosettes do not respond in the rosette inhibition test and the effect of EPF is undetectable. In addition, agents which may alter the lymphocyte integrity morphologically or metabolically, e.g. treatment with neuraminidase, addition of azide to the incubation medium, use of heparin with preservatives, residues of detergents and washing reagents on glassware or even the presence of magnesium or calcium ions have been found to disturb the test system (Morton et al., 1982b).

An improvement of the rosette inhibition test for the detection of EPF is hoped for from the use of monoclonal antibodies instead of ALS (Tinneberg et al, 1984; Rolfe et al., 1984). This would not only enable different groups to establish the test and compare results, but would also facilitate investigations into the interactions between EPF and lymphocytes. Rolfe et al. (1984) recommend a modified rosette inhibition test so that EPF in sera from different species can be detected with mouse lymphocytes. Interference of the assay which results from heterologous proteins is prevented by the introduction of an ion exchange chromatography step. The mouse system has several advantages as it is reported to be extremely sensitive, controlled populations of lymphocytes are readily prepared and a number of monoclonal antibodies against lymphocytes are available.

By strictly following the instructions described to be successful for the demonstration of EPF (Morton et al., 1982b and 1983), we tried a partial validation of the rosette inhibition test for the detection of EPF-activity in pig serum (Koch and Ellendorff, 1985a). Lymphocytes were collected from a total of 25 castrated pigs. The cells of all donors formed rosettes with sheep red blood cells (average of 11.3% rosette-forming lymphocytes). There was no difference for the percentage of spontaneous rosettes after preincubation of lymphocytes in non-pregnancy or pregnancy serum.

After the preincubation of lymphocytes in non-pregnancy sera and using a standard-ALS for all tests, a mean RIT of 10.7 was determined. The upper limit of the 95%-confidence interval was calculated to be 11.2. Thus, all titres > 12 were defined as positive for EPF-activity. However, employing this definition of EPF-activity, positive titres were also detected in 8.6% non-pregnancy sera (n = 70). The incubation in pregnancy serum caused a significant increase in the mean RIT raising it to 14.5 (n = 205). EPF-positive titres were found in the sera of only 114 (55.6%) of the pregnant pigs.

Unlike the formation of spontaneous rosettes the lymphocytes of some donors never, and those of the remaining animals only inconsistently induced an elevated RIT after preincubation in serum of pregnant pigs. To exclude this failure of the test system, known EPF-positive and EPF-negative control samples were included in each test. Data were only utilized if these samples gave the expected results. Nevertheless, after repeated assays of a pregnancy serum, EPF-positive titres were detected in only 25 of 30 tests, although both control samples did not indicate any fluctuation of the rosette inhibition test. Thus, failures of the rosette inhibition test are not always detectable by the inclusion of controls as stated by Morton et al. (1982b). We failed to identify the cause of this inconsistency, and similar difficulties have been reported by other groups (Paisley et al., 1982, Smart et al., 1982). While all points mentioned before may explain a complete breakdown of the test system, they do not offer an explanation for the failure of only a few samples within an otherwise valid test.

To improve the reliability of the rosette inhibition test we tried a modification. All EPF-negative samples were reanalysed up to a total of three times, independently of their origin from pregnant or non-pregnant animals, and provided no EPF-activity was found in a second test. Following this modification a correct pregnancy diagnosis occurred in 88.7% of all cases (n = 275). A total of 8.6% (n = 70) false positive and 12.4% (n = 205) false negative results were obtained during the whole gestation period. In general, the rosette inhibition test proved to be more reliable for the analysis of non-pregnancy sera when compared with pregnancy serum samples. The reproducibility of the RIT with non-pregnancy serum was considered good (inter-assay coefficient of variation 6.5%, n = 20), though it was less satisfactory with pregnancy serum (inter-assay coefficient of variation 23.8%, n = 30).

In a "blind" study with 20 human serum samples Smart et al. (1982) reported a 80% accuracy of pregnancy diagnosis. Two samples failed to correlate with the pregnancy or non-pregnancy status. From 25 tested non-pregnancy sera, 23 (93%) gave an RIT within the non-pregnancy range, while 21 of 25 (84%) pregnant sera gave an EPF-positive RIT.

Morton et al. (1979b) found an EPF-negative result in only 2 of 40 serum samples collected from pregnant ewes during the first month of pregnancy. Both samples were taken from the same animal at different times. All samples from non-pregnant animals (n = 40) had titres within the non-pregnancy range.

PRODUCTION OF EPF

Time course during pregnancy

A number of studies have been carried out to follow the time course of EPF-production during pregnancy in different species. In the mouse the first increase of the RIT is detectable within 4-6 h after mating (Morton et al., 1976). In a group of 7 mated sheep a significantly elevated titre was observed within 24 h after mating (Morton et al., 1979a). However, pregnancy was diagnosed in these animals only by the absence of the following behavioural oestrus. Confirmed pregnancies were detected in sheep within 72 h after mating. In one pregnant ewe an increased RIT was observed 6 h post coitum (p.c.) (Morton et al., 1979b). Preliminary investigations suggest that in humans fertilization may be detectable within 48 h, although a confirmation of pregnancy by established clinical methods is missing in all reported cases (Morton et al., 1977; Rolfe, 1982).

While Grewal et al. (1981) were unable to detect an enhanced RIT with pig serum before week 3 to 4 of pregnancy, EPF-activity has been detected in this species much earlier, provided the serum samples were dialysed beforehand (Morton et al., 1983; see below). The time course of EPF-production in the pig is shown in Fig. 3. As early as 4 h after mating, serum samples from two pigs induced an EPF-positive titre. By

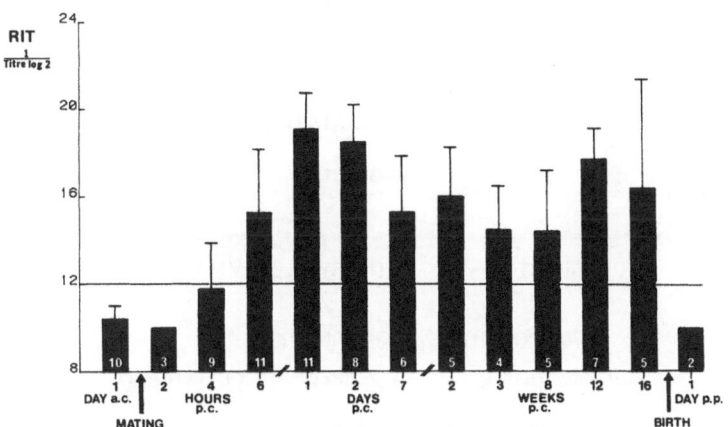

Fig. 3   Time course of EPF-production in the pig.   The bars
represent the mean RIT (± S.E.M.) for the number of serum
samples indicated.   Samples were collected from the same animals
before insemination, at various times during pregnancy and after
parturition.   An RIT > 12 indicates the presence of EPF-
activity (from Koch et al., 1982).

24 h p.c. EPF-activity was observed in all animals which later were found
to be pregnant (Koch et al., 1982; Morton et al., 1983).

The serum EPF-activity in the mouse disappears some 4 to 6 days
before birth (Morton et al., 1976).   Similarly, in sheep and humans the
elevated RIT drops to the non-pregnancy range during the second half or
the last third of pregnancy (Morton et al., 1977; Morton et al., 1979b;
Smart et al., 1982).   The pattern of EPF-production in the pig is completely different as EPF-positive titres are present virtually to the end
of pregnancy with lower RITs during mid-pregnancy (Morton et al., 1983;
Koch et al., 1985a).   The originally suggested biphasic production of EPF
has now been shown to be a polyphasic phenomenon with regular fluctuations
during the first half of pregnancy, which resemble the physiological
oestrous cycle (Koch and Ellendorff, 1985a).   Peaks of RITs were always
followed by a marked decline of EPF-activity around days 20, 40 and 60 of
gestation.   Especially at these pregnancy stages, but generally between
days 35 and 60 p.c., an increased number of EPF-negative animals was
observed.   During the second half of pregnancy the proportion of
EPF-negative animals as well as the fluctuation of RITs were diminished.

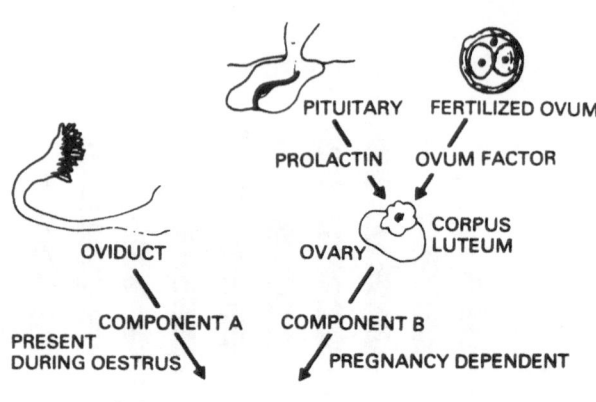

Fig. 4   Hypothetical model of EPF-production in the mouse.
For details see text (after Morton et al., 1982a).

Mode and site of production

Using the mouse as an experimental model to determine the site of
EPF-production, Morton et al., (1980) collected evidence that EPF consists
of two separate components, called EPF-A and EPF-B. EPF-A seems to be
produced by the oviduct during oestrus and pregnancy and EPF-B by the
ovary during pregnancy (Fig. 4).

While the mechanism controlling the production of EPF-A is not yet
known, in vitro cultures of different mouse organs, incubated alone or in
various combinations, have revealed that the production of EPF-B can be
induced in ovaries from animals at oestrus or dioestrus in the presence of
both fertilized ova and pituitaries from mice at oestrus and dioestrus.
After replacement of the pituitary in the culture system by prolactin, FSH
and LH, only prolactin was able to stimulate the production of EPF in
non-gravid ovaries. Prolactin, therefore, has been regarded as the
crucial pituitary factor. This assumption has further been strengthened
by in vivo experiments, as it has been demonstrated that EPF disappears
within 48 h after hypophysectomy, but the production can be maintained by
two daily injections of 10 μg prolactin (Morton et al., 1982a).

The second precondition for the production of EPF, a signal from the
fertilized ovum, was initially postulated during the above described in

vitro studies (Morton et al., 1980). In the meantime the existence of a
humoral factor has also been established in vivo. After the transfer of
fertilized and unfertilized ova in pseudopregnant mice, EPF-activity was
only detected in those animals which had received viable embryos (Morton
et al., 1982a). Nancarrow et al. (1981) infused 39,000 x g supernatants
of frozen-thawed fertilized and unfertilized mouse ova into the lumen of
the oviduct of oestrous sheep ipsilateral to the site of ovulation. While
the supernatant of unfertilized ova did not influence the serum titre, the
infusate prepared from fertilized ova increased the RIT to the pregnancy
range at 1 h. The titre was further elevated at 3 h but had returned to
non-pregnancy values at 22 h. The medium in which the ova were fertilized
caused a small rise at 2.5 and 3 h. In oestrous mice, on the other hand,
EPF-production could be stimulated by the injection of culture medium from
human and pig embryos (Morton et al., 1982a). This indicates that the
action of the embryonic signal, called Zygotin (Nancarrow et al., 1981) or
Ovum Factor (OF) (Cavanagh et al., 1982), is not species-specific.

Using an indirect assay for mouse OF based on the detection of EPF
Cavanagh et al. (1982) showed that the embryonic signal is released from
virtually the time of penetration of the ovum by a spermatozoon, until at
least the blastocyst stage. The detection of EPF-activity in a sheep only
1 h after the estimated time of ovulation and fertilization (Nancarrow et
al., 1981) confirms the extremely early release of OF and supports the
idea that OF is already present in the unfertilized ova (Cavanagh et al.,
1982). The supposition that OF may be a sperm product which is released
following the interaction with the ovum seems less likely, as the
secretion of OF could also be stimulated by the incubation of unfertilized
mouse ova in medium without calcium and magnesium ions or medium
containing hyaluronidase (Cavanagh et al., 1982). Both these conditions
are known to be able to activate mouse ova parthenogenetically (Kaufmann,
1978). However, as the mechanical disruption of the vitelline membrane
did not initiate the secretion of OF, a specific releasing mechanism
appears to be involved (Cavanagh et al., 1982). A preliminary
characterization of mouse OF showed that it exists in multiple molecular
weight (MW) forms of approximately 160,000, 2,800 and 1,500. Injection of
both the small or the large MW form of OF increased the serum RIT of
oestrous mice within 2.5 h post injection, while the elution buffer had no
effect (Cavanagh et al., 1982).

The validation of the above outlined mechanism of EPF-production has to be established for other species. Even in the mouse the results show some contradictions. Pituitaries taken from pregnant mice 24 h p.c. have been found to initiate the synthesis of EPF-B (Morton et al., 1980). This suggests that the hypophysis, once stimulated, does not depend on the permanent support of an embryonic signal and consequently the continuous presence of a viable conceptus to produce the active compound. Inconsistent with the above model is also the demonstration of EPF-like activity in day 2 and 3 human embryo cultures (Smart et al., 1981). In addition, the capacitation and insemination media were EPF-positive, while control media were negative in the rosette inhibition test. The factor in capacitation medium was shown to be dialyzable and thus different from EPF. Insemination and growth media, however, contained non-dialyzable molecules (Roberts and Smart, 1983).

There is some evidence that the embryo from the blastocyst stage onwards itself is a source of EPF, as EPF-activity is then detectable in culture medium (Morton et al., 1982a). If pregnancy was maintained in ovariectomized mice by progesterone replacement therapy, EPF-activity was found only in those animals which carried embryos at least at the blastocyst stage. Moreover, ovaries and oviducts of pregnant mice beyond day 6 could no longer be stimulated to produce EPF in vitro, although serum EPF-activity is usually detected until day 14 to 16 of gestation. Thus, a two-phase mechanism of EPF-production has been implicated. While the oviduct and the ovary maintain EPF-production until about day 6 of pregnancy, a second source develops around day 4 in the embryo and continues production until the last trimester (Morton et al., 1982a). Since EPF-activity is still detectable in the sheep fetus after it disappears from the ewe serum, a similar mode of EPF-production might function in the sheep. The identification of EPF-like material in sheep placenta cultures possibly points to the source of fetal EPF (Morton et al., 1982a). Also, in the pig there is some indication that the fetuses are involved in EPF-production (Koch, 1983).

Relationship of EPF to the presence of a viable conceptus

The detection of EPF-activity within hours after mating or insemination is thought to be specific for fertilization and the continuous presence of a viable conceptus (Morton et al., 1982a). This contention is based on observations that neither the presence of spermatozoa in the genital tract of salpingectomized or tubal ligated female mice (Morton et al., 1976), sheep (Morton et al., 1979) and humans (Smart et al., 1982), nor mating of intact female mice with a vasectomized buck (Morton et al., 1976) increased the serum-RIT of a standard ALS. But since the oviduct seems to be involved in EPF-production these experiments might not be relevant.

After the transfer of fertilized and unfertilized ova in synchronized mice, EPF-activity was only detected in animals which had received viable embryos (Morton et al., 1982a). Similar experiments were done in sheep (Nancarrow et al., 1981) and pigs (Koch, Niemann and Ellendorff, unpublished). Following the transfer of 14 or 15 embryos into 2 sows an increased serum-RIT was evident within 2-4 h. The failure of further development of the embryos was coincident with a disappearance of EPF-activity at 24 and 28 h in both recipients.

In pregnant sheep (day 19 to 20) the raised serum-RITs fell to the non-pregnancy range within 48 h after interruption of pregnancy by induction of luteolysis or surgical removal of embryos (Nancarrow et al., 1979). At 24 h after flushing the uteri of 8 pregnant sows (day 3 to 6) an EPF-positive titre was found in only one animal but at a markedly reduced value (Koch, Niemann and Ellendorff, unpublished).

In humans (Morton et al., 1977; Roberts and Smart, 1983; Rolfe, 1982), as in sheep (Morton et al., 1979a) and pigs (Koch et al., 1983), EPF-activity is frequently observed immediately post copulation but disappears within two to three weeks. These findings suggest a high rate of fertilization combined with a considerable rate of early embryonic failure. Rolfe (1982) detected EPF-activity in 18 of 26 cycles (69%) in 11 women. In 14 cycles EPF-activity disappeared within 14 days. Successful pregnancy was maintained in two cases, while the other two ended in miscarriages which were preceded by a disappearance of EPF. β-hCG was demonstrated in serum samples of all four women when tested more than 14 days after fertilization. Similarly, Smart et al. (1983, cited by Roberts and Smart, 1983) discovered a preimplantation loss of 50%. Of 21 cycles from 18 women studied, 14 had detectable EPF-activity. In 9 of these positive cycles β-hCG was also detectable.

Since in farm animals the early confirmation of fertilization by demonstration of chorionic gonadotrophins is not possible, we chose a different approach to answer the question whether EPF-production in the pig may indeed solely be stimulated by fertilization or may also be induced by sterile mating. For this purpose ten pigs were mated with a vasectomized boar and serum samples were collected immediately before copulation and 24, 48 and 72 h thereafter. Surprisingly, in the serum of 6 of 9 animals EPF-activity was detected at 24 h, while at 48 and 72 h EPF-positive results were obtained with decreasing titres in only one sow. Further experiments strongly supported the hypothesis that EPF-activity is primarily due to the presence of seminal plasma within the female genital tract (Koch and Ellendorff, 1985b). Non-specific activation of ova to release OF which in turn induces the transitory production of EPF or the resorption of an EPF-like seminal plasma constituent into the female blood circulation are considered as possible explanations. Bearing in mind the large ejaculate volume of the boar, it is conceivable that the observed effect of sterile mating is exclusively attributable to the pig. However, the detection of EPF-like activity in serum of men with certain testicular germ cell tumours has already shown that this phenomenon is not restricted to the female sex (Rolfe et al., 1983b).

As these investigations show the demonstration of EPF-activity during the first 24 h after mating is not necessarily pregnancy-specific in the pig. Nevertheless, the sustained production of EPF has also been found in this species to depend on the presence of viable conceptuses. Interestingly, for the polytocous pig more than two embryos seem to be required to signal EPF-production, while the level of rosette inhibition is closely correlated with the number of fertilized ova (Koch, Niemann and Ellendorff, unpublished).

BIOLOGICAL FUNCTION OF EPF

Since its discovery the synergistic effect of EPF and immuno-suppressive ALS has been related to an alteration of immunological reactivity during pregnancy. As the rosette inhibition test involves T-lymphocytes it has been suggested that mainly the cellular immune reaction is changed by EPF. To test this hypothesis in vivo Morton et al. (1982a) chose the cell mediated hypersensitivity as an experimental model. Employing the contact sensitivity against dinitrochlorobenzene (DNCB) Fabris (1973) had already shown that in mice the pregnancy status indeed diminishes the expression of the allergic reaction during a second contact but does not influence the induction of cell-mediated immunity. When non-pregnant mice were injected with EPF-containing material during priming with DNCB only a slight reduction of the cell-mediated hypersensitivity was observed. However, injection of EPF immediately before the second contact completely abolished the contact dermatitis, while the antibody production against sheep red blood cells was not influenced (Morton et al., 1982a).

Noonan et al. (1979) applied the technique of adoptive transfer of contact sensitivity against trinitrochlorobenzene (TNCB) to study the immunosuppressive effect of EPF. The indirect technique was chosen as the direct injection of sheep EPF had indicated that some effects were caused solely by heterologous protein. Lymphocytes of sensitized donors were incubated in serum or serum fractions of pregnant and non-pregnant sheep before the injection into syngeneic mice. With all samples of pregnant animals, in which EPF-activity had been detected, a significant reduction of the immune response was observed. These experiments suggest that EPF exerts an immunosuppressive effect by preventing antigen recognition by sensitized T-lymphocytes. Moreover, the effect of EPF seems not to be species-specific.

Immunosuppressive properties have been attributed to a number of serum factors, including blocking antibodies, hormones and pregnancy-related proteins. However, the clinical relevance of these substances for the protection of the semi-allogeneic conceptus against the maternal immune defence has to be established for most of them. This is especially true for EPF and the mechanism of its action is still far from clear. Nevertheless, if this function of EPF can be confirmed in further investigations then EPF might achieve importance not only for the detection of early embryonic mortality but also for the prevention of some of its causes.

BIOCHEMICAL CHARACTERIZATION OF EPF

Biochemical characteristics of EPF have been studied in mice (Clarke et al., 1978; Cavanagh, 1984), sheep (Clarke et al., 1980; Wilson et al., 1983; Wilson et al., 1984), pig (Morton et al., 1983) and humans (Rolfe et al., 1983a). Serum, urine, placentae and the mouse ovary-oviduct culture system have been used as sources of material for the isolation of EPF-active fractions.

Gel filtration studies suggest that EPF is a protein with a multiple molecular weight (MW) structure (Clarke et al., 1980). Results obtained by DEAE-cellulose chromatography show that EPF is present in the serum both free and bound to a carrier molecule which is a constituent of normal serum (Clarke et al., 1980). EFP bound to the carrier is predominant during the preimplantation period of pregnancy and has an apparent MW of $\geq$ 250,000. The free form prevails after implantation and is present in polymers with a basic MW of about 20,000. The same general characteristics have been observed in all species studied.

EPF from 24 h pregnant sheep can be further divided into two components by 40% ammonium sulphate (Clarke et al., 1980). Interestingly, the derived components appear to have their in vivo counterparts. The component soluble in 40% ammonium sulphate seems to be identical with EPF-A, while the insoluble substance corresponds with EPF-B (Clarke et al., 1980). EPF-A apparently binds to lymphocytes but will only adhere in the presence of EPF-B (Clarke et al., 1980; Morton et al., 1980). Both moieties of the EPF-complex are essential for the detection of EPF-activity by the rosette inhibition test. However, if EPF-A is present in excess after fertilization, as observed in the pig, it may block the binding of EPF and cause a negative assay. This problem can be overcome by the removal of excess EPF-A from serum samples by a dialysis step before the preincubation of lymphocytes (Morton et al., 1983).

Cavanagh (1984) in a three-step procedure involving immunoabsorption, electrofocusing, and gel filtration, isolated a homogeneous fraction from the mouse ovary-oviduct culture system which was responsible for 90% of the original EPF-activity. This fraction with pI 6.83 and MW of 21,000 could be further split by SDS treatment into 3 peptides of MW 10,500, 7,200 and 3,400, the first having activity equivalent to EPF-A, while the remaining two combined and had an activity resembling EPF-B.

Wilson et al. (1983) recently reported the isolation of EPF from serum of ewes between weeks 3 to 8 of gestation. Their procedure utilized molecular size exclusion techniques, ion exchange chromatography as well as high performance liquid chromatography and yielded two homogeneous active fractions with MW 20,000 and 67,000. The 20,000 polypeptide form appeared to represent the major form of EPF-active material. Further biochemical characterisation of both active polypeptides clearly distinguished them from the ovine hormones prolactin, placental lactogen and growth hormone (Wilson et al., 1984). However, these authors failed to further fractionate either of these polypeptides by ammonium sulphate. Because of the different biochemical behaviour of EPF-activity isolated from sheep at different stages of pregnancy and the possible involvement of different sources (see above), Clarke and Wilson (1982) have speculated on the existence of "early" and "late" forms of EPF.

CONCLUSION

As this review may have shown, EPF is a fascinating biological phenomenon possessing a number of exciting features which make it most interesting for both applied clinical investigations as well as basic research in reproductive physiology and immunology. However, at present EPF is detectable only by using the rosette inhibition test. The limitations of this assay system have been pointed out. It is cumbersome, tedious and an easy subject of disturbance. Of particular disadvantage is the indirect character of the procedure and its possible interference with other substances. As the rosette inhibition test is an almost complete bioassay which uses normal cells and sera, it is particularly susceptible to natural fluctuations. While the detection of EPF-activity by the rosette inhibition test may be possible for limited investigations under well-defined experimental conditions, it is neither useful for monitoring the isolation and purification of EPF nor for practical application such as for routine pregnancy diagnosis. Highest priority should be placed on the development of a reliable, sensitive, specific and rapid assay system. This might be achieved by analyzing the mechanism of rosette inhibition or the production of - preferably monoclonal - antibodies to EPF.

ACKNOWLEDGEMENTS

I wish to thank Dr R.B. Heap for his advice and criticism in preparing this review and Mrs J.A. Tickner for typing the manuscript.

REFERENCES
Ayalon, N. 1978. A review of embryonic mortality in cattle. J. Reprod. Fert., 54, 483-493.
Bach, J.F. and Antoine, B. 1968. In vitro detection of immunosuppressive activity of antilymphocyte sera. Nature, 217, 658-659.
Bach, J.F., Dardenne, M. and Fournier, C. 1969. In vitro evaluation of immunosuppressive drugs. Nature, 222, 998-999.
Binns, R.M. 1982. Organisation of the lymphoreticular system and lympho-

cyte markers in the pig. Vet. Immunol. Immunopathol., 3, 95-146.

Cavanagh, A.C. 1984. Production in vitro of mouse early pregnancy factor and purification to homogeneity. J. Reprod. Fert., 71, 581-592.

Cavanagh, A.C., Morton, H., Rolfe, B.E. and Gidley-Baird, A.A. 1982. Ovum factor: A first signal of pregnancy? Am J. reprod. Immunol., 2, 97-101.

Clarke, F.M., Morton, H. and Clunie, G.J.A. 1978. Detection and separation of two serum factors responsible for depression of lymphocyte activity in pregnancy. Clin. exp. Immunol., 32, 318-323.

Clarke, F.M., Morton, H., Rolfe, B.E. and Clunie, G.J.A. 1980. Partial characterisation of early pregnancy factor in the sheep. J. reprod. Immunol., 2, 151-162.

Clarke, F. and Wilson S. 1982. Biochemistry of early pregnancy factor. In "Pregnancy Proteins" (Ed. J.G. Grudzinskas, B. Teisner and M. Seppala). (Academic Press, Sydney, New York). pp. 407-412.

Cooper, D.W. and Aitken, R.J. 1981. Failure to detect altered rosette inhibition titres in human pregnancy serum. J. Reprod. Fert., 61, 241-245.

Edey, T.N. 1979. Embryo mortality. In "Sheep Breeding" (Ed. G.L. Tomes, D.E. Robertson and R.J. Lightfoot). (Butterworth, London, Sydney). pp. 315-333.

Fabris, N. 1973. Immunological reactivity during pregnancy in the mouse. Experientia, 29, 610-612.

Flint, A.P.F., Saunders, P.T.K. and Ziecik, A.J. 1982. Blastocyst-endometrium interactions and their significance in embryonic mortality. In "Control of Pig Reproduction" (Ed. D.J.A. Cole and G.R. Foxcroft). (Butterworth Scientific, London, Boston). pp. 253-275.

Grewal, A.S., Wallace, A.L.C., Pan, Y.S., Rigby, N.W. and Nancarrow, C.D. 1981. A serum factor in pregnant pigs detected by a rosette inhibition test. Proc. A. Mtg. Austral. Soc. Reprod. Biol., Christchurch 1981, p. 96 (Abstr.).

Kaufman, M.H. 1978. The experimental production of mammalian partheno-genetic embryos. In "Methods in Mamalian Reproduction" (Ed. J.C. Daniel Jr.). (Academic Press, New York, San Francisco). pp. 21-47.

Koch, E. 1983. Early Pregnancy Faktor - Aktivitaten beim Schwein. Vet. Med. Diss. Hannover.

Koch, E. and Ellendorff, F. 1985a. Prospects and limits of the rosette inhibition test to detect EPF-activity in the pig. J. Reprod. Fert. (in press).

Koch, E. and Ellendorff, F. 1985b. Detection of EPF-like activity after mating sows with a vasectomized boar. J. Reprod. Fert. (in press).

Koch, E., Morton, H. and Ellendorff, F. 1983. Early pregnancy factor: Biology and practical application. Br. vet. J., 139, 52-58.

Koch, E., Morton, H., Morton, D. and Ellendorff, F. 1982. Fruher Trachtigkeitsfaktor (EPF) beim Schwein. In "Physiologie und Pathologie der Fortpflanzung. 7. Veterinar-Humanmedizinische Tagung, Giessen 1982." (ed. K. Semm, H. Tillmann, W. Gehring and L. Mettler). pp. 168-170.

Marcus, Z.H., Hess, E.V., Herman, J.H., Troiano, P. and Freisheim, J. 1979. In vitro studies in reproductive immunology - 2. Demonstration of the inhibitory effect of male genetial tract constituents on PHA-stimulated mitogenesis and E-rosette formation of human lympho-cytes. J. reprod. Immunol., 1, 97-107.

Morton, H., Clunie, G.J.A. and Shaw, F.D. 1979b. A test for early pregnancy in sheep. Res. vet. Sci., 26, 261-262.

Morton, H., Hegh V. and Clunie, G.J.A. 1974. Immunosuppression detected in pregnant mice by rosette inhibition test. Nature, 249, 459-460.

Morton, H., Hegh, V. and Clunie, G.J.A. 1976. Studies of the rosette inhibition test in pregnant mice: Evidence for immunosuppression. Proc. R. Soc. Lond. B., 193, 413-419.

Morton, H., Morton, D.J. and Ellendorff, F. 1983. Appearance and characteristics of early pregnancy factor in the pig. J. Reprod. Fert., 68, 437-446.

Morton, H., Nancarrow, C.D., Scaramuzzi, R.J., Evison, B.M. and Clunie, G.J.A. 1979a. Detection of early pregnancy in sheep by the rosette inhibition test. J. Reprod. Fert., 56, 75-80.

Morton, H., Rolfe, B. and Cavanagh, A. 1982a. Early pregnancy factor: Biology and clinical significance. In "Pregnancy Proteins" (Ed. J.G. Grudzinskas, B. Teisner and M. Seppala). (Academic Press, Sydney, New York). pp. 391-405.

Morton, H., Rolfe, B., Clunie, G.J.A., Anderson, M.J. and Morrison, J. 1977. An early pregnancy factor detected in human serum by the rosette inhibition test. Lancet, I, 394-397.

Morton, H., Rolfe, B.E., McNeill, L., Clarke, P., Clarke, F.M. and Clunie G.J.A. 1980. Early pregnancy factor: Tissues involved in its production in the mouse. J. reprod. Immunol., 2, 73-82.

Morton, H. Tinneberg, H.R., Rolfe, B., Wolf, M. and Mettler, L. 1982b. Rosette inhibition test: A multicentre investigation of early pregnancy factor in humans. J. reprod. Immunol., 4, 251-261.

Nancarrow, C.D., Evison, B.M., Scaramuzzi, R.J. and Turnbull, K.E. 1979. Detection of induced death of embryos in sheep by the rosette inhibition test. J. Reprod. Fert., 57, 385-389.

Nancarrow, C.D., Wallace, A.L.C. and Grewal, A.S. 1981. The early pregnancy factor of sheep and cattle. J. Reprod. Fert. Suppl., 30, 191-199.

Noonan, F.P., Halliday, W.J., Morton, H. and Clunie, G.J.A. 1979. Early pregnancy factor is immunosuppressive. Nature, 278, 649-651.

Paisley, L.G., Davis, W.C., Anderson P.B. and Mickelsen, W.D. 1982. Detection of early pregnancy factor in swine: A need for dialogue. Theriogenology, 18, 393-401.

Roberts, T.K. and Smart, Y.C. 1983. Studies on human early pregnancy factor. In "Reproductive Immunology". (Ed. S. Isojima and W.D. Billington). (Elsevier, Amsterdam). Proc. 2nd Int. Cong. Reprod. Immunol., Kyoto, Japan, Aug. 17-20, 1983. pp. 157-169.

Rolfe, B.E. 1982. Detection of fetal wastage. Fert. Steril., 37, 655-660.

Rolfe, B., Cavanagh, A., Forde, C., Bastin, F., Chen, C. and Morton, H. 1984. Modified rosette inhibition test with mouse lymphocytes for detection of early pregnancy factor in human pregnancy serum. J. Immunol. Methods, 70, 1-11.

Rolfe, B.E., Morton, H., Cavanagh, A.C. and Gardiner, R.A. 1983b. Detection of an early pregnancy factor-like substance in sera of patients with testicular germ cell tumors. Am. J. reprod. Immunol., 3, 97-100.

Rolfe, B.E., Morton, H. and Clarke, F.M. 1983a. Early pregnancy factor is an immunosuppressive contaminant of commercial preparations of human chorionic gonadotrophin. Clin. exp. Immunol., 51, 45-52.

Smart, Y.C., Cripps, A.W., Clancy, R.L., Roberts, T.K., Lopata, A. and Shutt, D.A. 1981. Detection of an immunosuppressive factor in human preimplantation embryo cultures. Med. J. Aust., 1, 78-79.

Smart, Y.C., Roberts, T.K., Fraser, I.S., Cripps, A.W. and Clancy, R.L. 1982. Validation of the rosette inhibition test for the detection of early pregnancy in women. Fertil. Steril., 37, 779-785.

Stimson, W.H. 1976. Studies on the immunosuppressive properties of a pregnancy-associated alpha-macroglobulin. Clin. exp. Immunol., 25, 199-206.

Thomson, A.W., Milton, J.I., Campbell, D.M. and Horne, C.H.W. 1980. Rosette inhibition levels during early human gestation. J. reprod. Immunol., 2, 263-268.

Tinneberg, H.R., Staves, R.P. and Semm, K. 1984. Improvement of the rosette inhibition assay for the detection of early pregnancy factor in humans using the monoclonal antibody, anti-human-Lyt-3. Am. J. reprod. Immunol., 5, 151-156.

Whyte, A. and Heap, R.B. 1983. Early pregnancy factor. Nature, 304, 121-122.

Wilson, S., McCarthy, R. and Clarke, F. 1983. In search of early pregnancy factor: Isolation of active polypeptides from pregnant ewes' sera. J. reprod. Immunol., 5, 275-286.

Wilson, S., McCarthy, R. and Clarke, F. 1984. In search of early pregnancy factor: Characterisation of active polypeptides isolated from pregnant ewes' sera. J. reprod. Immunol., 6, 253-260.

UTERINE ENVIRONMENT IN EARLY PREGNANCY *

B. Fischer & H. M. Beier

Abteilung Anatomie und Reproduktionsbiologie
Medizinisch-Theoretische Institute
Medizinische Fakultät
RWTH Aachen
Melatener Straße 211
D-5100 Aachen
Federal Republic of Germany

ABSTRACT

Mammalian species have developed an extraordinary biochemical and endocrinological diversity to ensure reproduction. But in all species synchrony between maternal and embryonic development is obligatory to establish pregnancy. The composition of the uterine environment during the preimplantation period which represents an essential part of this synchrony will be reviewed.

The over-all quality of the uterine environment - the maternal "soil" - is predominantly a permissive one. It determines reproductive performance of the female but has no inductive function on the genetically autonomous conceptus (quality of the "seed").

Analogous to embryonic development uterine secretions change quantitatively and qualitatively during early pregnancy. Among uterine secretory components, the proteins have been investigated most extensively and in many species. Most are of blood serum origin. But uterine proteins synthesized de novo by the endometrium also contribute - in some species to a considerable amount - to uterine luminal fluid during the preimplantational period. Typical protein pattern changes have been demonstrated as well as variations in enzyme activities. In some species pregnancy specific proteins have been identified representing embryonic-maternal interaction in early pregnancy. Uterine secretory activity is controlled by maternal hormones. In the rabbit, it has been shown that hormones dispose a complete inherent endometrial secretory programme.

Synchrony is defined by the hormonally controlled composition of the uterine luminal fluid in total (and not by single components) and by the cyclic changes occuring during the preimplantation period. An inadequate uterine milieu provokes embryonic mortality which is of importance in humans as well as in farm animals.

* Dedicated to C.E. Adams (1926-1984)

INTRODUCTION

During the preimplantation period the uterine environment undergoes typical changes which can be defined morphologically, biochemically, enzymologically, endocrinologically, and biologically. The last aspect reflects the difficulty in adequately defining the quality of the uterine environment for the developing conceptus by one criterion. Only the biological proof - implantation and normal postimplantation development to the birth of healthy offspring - reveals the sufficiency of the uterine environment.

In the following presentation we have tried to define the uterine environment by its biochemical composition and endocrine regulation, demonstrating its decisive effect on embryo development, and trying to offer possible reasons for embryonic mortality during the preimplantation period.

SIGNIFICANCE OF UTERINE ENVIRONMENT

The uterine environment serves various functions in mammals.

It

- allows fertilization by enabling spermatozoa to ascend to the site of fertilization in the oviduct
- provides adequate nutrients for the different stages of embryonic development from arrival in the uterus after tubal transport to implantation
- maintains an appropriate milieu for the physico-chemical integrity of embryonic structure (e.g. demands of osmolarity)
- fulfills immunological requirements (both immunosuppressive and antibacterial requirements (see: Bazer & Roberts 1983, Strzemienski et al. 1984)).

The general quality of the uterine environment is thereby a permissive one. It must guarantee a milieu which offers optimal developmental conditions for the embryo but has no inductive functions, i.e. in determining the developmental potency of the conceptus. The latter is established by the embryonic genome, however, the embryo is dependent on an appropriate uterine environment in order to realize its inherent genetic

programme. The importance of the uterine environment is reflected in the reproductive performance of the mother.

Uterine secretions vary cyclically according to different demands of the developing embryo. These changes involve both
quantitative and qualitative secretory activity and differ
considerably among species. They are controlled by maternal
hormones as discussed later. We think that these changes represent the underlying mechanism of synchrony between maternal
and embryonic development. Synchrony is a basic principle in
early mammalian reproduction and can be conclusively demonstrated by embryo transfer experiments. They yield highest
success rates if synchronization has been precise. This was
first demonstrated in detail in the rabbit by Chang (1950).
The biological proof of uterine secretions as a principle of
the synchrony was initially shown by Beier et al. (1972).
Rabbit embryos were transferred into 'desynchronized' recipients 4 days out of phase. The desynchronization was induced
by post-coital estrogen treatment and verified by uterine
protein patterns (Beier 1973, 1974). This was later confirmed
by Adams (1974) using a different experimental schedule.

For successful reproduction in cows and sheep, e.g., the disparity between embryonic and uterine development should not
exceed $\pm$ 1 and $\pm$ 2 days, respectively (Rowson & Moor 1966,
Rowson et al. 1969, Rowson 1971, Gordon 1982, Moore 1982).

DEMANDS OF THE EMBRYO

Cleavage stage embryos are more independent from environmental conditions than blastocysts. In the rabbit, e.g., environmental changes are not as pronounced in the oviduct compared with the uterus (Beier 1974). Blastocysts, however, develop normally for a longer time only in an
(1) adequate and
(2) uterine milieu.
It has been conclusively demonstrated in previous years that
embryos restricted to the oviductal environment fail to develop beyond blastocyst formation in most species (Adams 1958,
Kirby 1962, Murray et al. 1971, Fischer et al. 1981). In the

human, however, embryos can develop and even implant in the
Fallopian tube, as shown by ectopic pregnancies.

In cattle, sheep, and pigs blastocysts spend a prolonged period in the uterus before implantation and expand remarkably
after hatching from the zona pellucida (Cook & Hunter 1978,
Bazer & Roberts 1983). During the preimplantation period blastocysts have a high metabolic activity which requires sufficient
supply of precursors from the uterus (Surani 1977). But besides nutritional supply other requirements must also be met
by the uterine environment and attention has to be paid to its
influence, possibly selective, on the various embryonic compartments and developmental stages. A common system for investigating these questions is culture of embryos in vitro.
Culturing rabbit 2-c-stages in a protein-free medium leads to
   development up to the 16-c-stage, whereas morulae do not
develop to the blastocyst stage in such a medium (Ströbele-
-Müller 1984). These findings further demonstrate the different environmental demands of cleavage stage embryos and blastocysts.

In culture media containing an excess of nutrients notable
alterations in normal development can be observed after blastocyst formation in the rabbit. First, the development of the
embryoblast cells does not keep up with trophoblast growth,
i.e. while trophoblast cells still increase in cell number
embryoblast cells are no longer viable. Thus, these two embryonic cell lines differ in environmental demands (Beier 1979,
Beier et al. 1983). Secondly, the over-all rate of development
is progressively retarded with time in in vitro culture, i.e.
the longer the embryo suffers inadequate environmental conditions. The retardation is even more pronounced with advanced
developmental stage of the embryo. 2-c-rabbit embryos reached
a developmental stage after 4 days in vitro corresponding to
a 3 day old control embryo while morulae did not reach half
the diameter compared with in vivo controls after this time in
vitro (Ströbele-Müller 1984). Furthermore, extracellular as
well as cellular embryonic compartments do have specific needs.
Rabbit blastocyst coverings like the zona pellucida do not permit normal expansion of blastocysts (Beier 1974) and do not

undergo the physiological preimplantation transformation in
vitro (Fischer et al. 1982, 1983) or if trapped in the oviduct
by post-coital tubal ligation (Adams 1958, Fischer et al. 1981).
The zona pellucida cannot be dissolved under non-physiological
conditions. This physiological process coincides with the onset
of blastocyst expansion.

But individual blastocysts can adapt to suboptimal environmental conditions or to milieu changes. In rabbits, for example,
transfers of embryos cultured for 24 h to foster mothers resulted in an implantation rate of about 50%. All embryos transferred experienced the same environmental conditions but only
half survived. Non-physiological changes such as the incomplete
transformation of the rabbit blastocyst coverings was found to
be reversible following transfer to an adequate milieu. Another
example of the adaption of embryos to an inadequate uterine
milieu is the blastocyst diapause or quiescence (see later).
The metabolic activity of the blastocyst is resumed within
some hours if sufficient uterine secretion is restored (see
Surani 1977, Aitken 1982).

Up to now we have heard that a sufficient uterine environment
in early pregnancy should be synchronous in quantitative and
qualitative terms to permit normal embryonic development.

Uterine luminal fluid consists of energy substrates, ions, vitamins, amino acids, peptides, hormones, prostaglandins, enzymes, serum proteins, and uterine proteins. The latter in particular have been extensively investigated in many species.

Uterine flushing components can permeate the embryo (Surani
1977, Bazer & Roberts 1983). The predominating progestational
uterine protein in the rabbit, uteroglobin, can be detected in
cellular compartments as well as in blastocyst fluid (Beier &
Maurer 1975). But this interaction is not a one-way pathway.
Just recently, Godkin et al. (1984) were able to demonstrate
that a secretory protein of the sheep blastocyst selectively
stimulates synthesis of 6 polypeptides by the uterine endometrium.

UTERINE PROTEINS

In the species studied so far, most luminal uterine proteins are of blood serum origin. But uterine synthesized proteins contribute - in species like the pig and the rabbit to
a remarkable amount - to luminal fluid during early pregnancy.
In the rabbit, e.g., there is no accumulation of lower molecular weight serum protein in uterine flushings. We find a disparity in the relative amounts of blood serum proteins in uterine secretions so that we can postulate a selective transsu-
dation and not a simple sieve effect (Beier 1978). For example,
α_2-macroglobulin with a molecular weight of 820000 d is present in rabbit uterine flushings whilst haptoglobin or cerulo-
plasmin with molecular weights of approx. 100000 d and 132000 d,
respectively, are not. The albumin/globulin ratio of rabbit
blood serum is 1.32; in a day 6 p.c. rabbit uterine rinse, 0.15.

In contrast to uterine proteins, follicular fluid is released
into the antrum by a pure sieve effect leading to a lower concentration of high molecular weight serum proteins.

A synchronous uterine environment was defined earlier by quantitative and qualitative changes in the uterine composition
during early pregnancy.
Quantitatively, total protein concentration generally increases
with approaching time of implantation. Under qualitative aspects, uterine specific proteins should be mentioned first.
They are known to be synthesized in many species, in laboratory
rodents as well as in farm animals like the cow (Roberts &
Parker 1974a, Laster 1977, Dixon & Gibbons 1979), gilt (Squire
et al. 1972, Roberts & Bazer 1980, Bazer & Roberts 1983), ewe
(Roberts et al. 1976), and mare (Zavy et al. 1982b). These proteins are solely synthesized by the uterine endometrium and
contribute in different amounts to the uterine luminal proteins.
De novo synthesis was demonstrated by incorporation of uterine
administered radioactive labelled precursors into uterine proteins (Pratt 1977, Surani 1977, Fishel 1980, Mulholland &
Villee 1984). In the case of uteroglobin, the endometrial m-RNA
has been isolated and the amino acid sequence analyzed (Beato
& Rungger 1975, Bullock et al. 1976, Ponstingl et al. 1978,

Atger et al. 1980).

In some species and for some proteins a critical evaluation
was suggested by Martin (1984) and Milligan & Martin (1984).Our
knowledge on uterine proteins is based on various biochemical
and immunochemical analyses. A continuous synthesis and secretion might thereby be missed, as well as proteins present in
meagre amounts, labile compounds, or small peptides. By da-
maging the endometrium during flushing procedure artefacts
might be created, especially in progestational uteri at the
time of blastocyst implantation and in species with closure
of the uterine lumen such as the mouse, rat, and hamster
(Martin 1984, Milligan & Martin 1984).

Secondly, in regard to qualitative variations characteristic
protein pattern changes during the preimplantation period
should be emphasized. These changes have been shown to occur
in small rodents (rat (Surani 1976), mouse (Aitken 1977)) and
appear predominantly in the macromolecular region where whole
'families' of electrophoretic bands vary between different
progestational stages (Surani 1976). But traces of many of
these proteins can be demonstrated in other cyclic stages so
that quantitative rather than qualitative changes seem to cha-
racterize protein secretion patterns during early pregnancy
(Aitken 1982).

A fascinating model for studying the importance of uterine
components on embryonic development is their supplementation
of in vitro culture systems (El-Banna & Daniel 1972a, b,
Maurer & Beier 1976). We are still working on optimizing this
model but face technical and methodological problems including
furnishing sufficient amounts of uterine proteins, lack of biological activity following lyophilization, or loss of the low
molecular components by other techniques. The results obtained
so far demonstrate a beneficial effect of uterine components
in vitro expecially on embryoblast cells as shown by ultrastructural integrity (Beier et al. 1983).

ENZYMES
    Much work has been invested in recent years in studying

the pattern and significance of enzymes contributing to uterine
fluid. In early pregnancy, typical variations in enzyme activity are found which are mainly discussed in the context of
processing nutrients for the conceptus or initiation of implantation (Murdoch & White 1968, Murdoch 1970, Kirchner et al.
1971, Roberts & Parker 1974b, Denker 1977, Murdoch & O'Shea
1977, 1978, Murdoch et al. 1978, Denker 1982, Fazleabas et al.
1982, Peplow 1982, Zavy et al. 1982a, Bazer & Roberts 1983,
Zavy et al. 1984). Most likely, enzymes perform metabolic
transformations concerned with the nutrition of the embryo
during the preimplantation period, a period of considerable
metabolic activity. But the nature and mechanism of action of
many of the enzymes still remains obscure. Species differences
must be taken into account. The most challenging question is
the mode of regulation of enzyme activity, whether it is more
dependent on the maternal or embryonic system.

PREGNANCY SPECIFIC PROTEINS

    Uteroglobin is detectable in pregnant, pseudopregnant,
and non-pregnant rabbits, in different amounts and different
time patterns. In the rat changes in the uterine luminal protein occur in pregnant and non-pregnant luteal phase females
(Surani 1977).
In contrast, in the uterine flushings from cows Laster (1977)
demonstrated a pregnancy specific uterine protein of molecular weight of 50 - 60000 d. Quantitatively, this protein
accounted for about 1% or less of the uterine endometrial protein. Similarly, Roberts et al. (1976) report pregnancy specific proteins occuring about time of embryo attachment to the
endometrium in pregnant ewes.
Investigations on embryonic signals in early pregnancy revealed that products released by the pig or sheep conceptus in
utero - either hormones or proteins - led to the synchronous
secretion of proteins from endometrium epithelium (Heap 1980,
Heap et al. 1982, Bazer & Roberts 1983, Godkin et al. 1984).
Therefore, pregnancy specific proteins represent an important
link in embryo-maternal interaction and important markers for

further investigations on maternal recognition of pregnancy
and prevention of corpus luteum regression.

REGULATION OF UTERINE SECRETION

An excellent tool for studying embryo-maternal interactions is the quiescent or diapause blastocyst. Diapause can be
seasonally, lactationally, or experimentally (by ovariectomy)
initiated leading to impairment of uterine secretory and embryonic metabolic activity in species like the rat (see Surani
1977), mouse (see Aitken 1982), roe deer (Aitken 1974), mink
(Enders & Enders 1963), fur seal (Daniel 1971).
Resumption of synthesis and secretion of luminal proteins by
substitution of hormones to the ovariectomized female or resto-
ration of metabolic activity (DNA-, RNA-, protein synthesis,
glucose utilization, CO_2 production, cell division) by transfer of the embryo to an appropriate environment occurs within
some hours (see Surani 1977, Aitken 1982).
This model suggests a direct dependency of the embryonic development on
- uterine secretory activity and
- maternal hormones.

Since embryos can develop in extrauterine sites regardless of
the endocrine status of the mother (Kirby 1969) we can con-
clude that maternal hormones directly influence   u t e r i n e
embryonic development. The source of hormones may be the ovaries and/or higher centres like the pituitary, but could also
include the conceptus itself (see Heap et al. 1979).

In different species different hormonal requirements for uterine secretion are known. For example, whereas pigs and rabbits
require only progesterone for uterine protein synthesis and implantation, rats and mice additionally require estrogens.

The hormonal effects on morphology and metabolic activity of
endometrial cells are receptor-mediated. A selective distribution of steroid receptors among different cell types should
be noted.
The initiation of an inherent endometrial secretory programme

by hormones as an important regulatory mechanism in maternal-
-embryonic interaction was shown by Beier et al. in 1972 (see
Beier 1974). By post-coital estrogen application he was able
to delay uterine secretion in the rabbit for about 4 days.
After this time the normal preimplantation secretory pattern
started to demonstrate a kind of self-regulating
programme in the rabbit, obviously generated by post-ovulatory
endocrine commencement.

In 1981, we reported on advanced secretion of uteroglobin
following superovulation by gonadotropin stimulation in the
rabbit (Delbos et al. 1981). These findings lend further support to the susceptibility of the uterine secretions to exogenous or endogenous hormones in creating asynchronous conditions.

RELEVANCE OF UTERINE ENVIRONMENT TO REPRODUCTION

    In commenting on hormone action on the uterine environ-
ment and early embryonic development it should be mentioned
that injections of estrogens and progesterone into gilts led
to an increase in uterine secretory activity but litter size
and weight of embryos were not affected (Cook & Hunter 1978).

Hormones act indirectly via uterine secretions on embryonic
development or - in other terms - maternal hormonal imbalances
lead to a suboptimal or detrimental uterine environment.
A majority of embryonic loss occurs during the pre-and periimplantation period . In the human, failure of fertilization or
mortality during advanced stages of pregnancy is low compared
with the rate of pre- and periimplantation abortions (Boué
et al. 1975, Lauritsen 1982, Garcia 1983, Leeton et al. 1983).

Ayalon (1978) reports on the composition of endometrial secretions from normal and repeat-breeder cows. In accordance with
other authors he found differences in the concentrations of
ions, energy substrates, and proteins between these groups and
remarkable differences in the ion content between cows with
normal and abnormal conceptuses.
These observations strongly support the view that key information for defining embryonic mortality in early pregnancy can

be expected from further investigations on the uterine environment in the pre- and periimplantation period. Many components other than proteins - which have mainly been reviewed in this paper - define synchrony. All must be kept in mind when claiming an adequate uterine environment endowing optimal support to embryonic development in early pregnancy.

REFERENCES

Adams, C.E. (1958) Egg development in the rabbit: the influ-
ence of post-coital ligation of the uterine tube and of
ovariectomy.
J.Endocrin. 16, 283-293
Adams, C.E. (1974) Asynchronous egg transfer in the rabbit.
J.Reprod.Fert. 35, 613-614
Aitken, R.J. (1974) Delayed implantation in roe deer (Capreo-
lus capreolus).
J.Reprod.Fert. 39, 225-233
Aitken, R.J. (1977) Changes in the protein content of mouse
uterine flushings during normal pregnancy and delayed im-
plantation and following ovariectomy and oestradiol admi-
nistration.
J.Reprod.Fert. 50, 29-36
Aitken, R.J. (1982) The control of blastocyst activity. In
"Proteins and steroids in early pregnancy" (Ed. H.M. Beier,
P. Karlson) (Springer-Verlag, Berlin, Heidelberg, New York)
pp. 233-243
Atger, M., Mornon, J.P., Savouret, J.F., Loosfelt, H.,
Fridlansky, F. & Milgrom, F.C. (1980) Uteroglobin: a model
for the study of the mechanism of action of steroid hor-
mones. In "Steroid induced uterine proteins" (Ed. M. Beato)
(Elsevier/North-Holland Biomedical Press, Amsterdam, New
York, Oxford) pp. 341-350
Ayalon, N. (1978) A review of embryonic mortality in cattle.
J.Reprod.Fert. 54, 483-493
Bazer, F.W. & Roberts, R.M. (1983) Biochemical aspects of con-
ceptus-endometrial interactions.
J.exp.Zool. 228, 373-383
Beato, M. & Rungger, D. (1975) Translation of the messenger
RNA for rabbit uteroglobin in Xenopus oocytes.
FEBS Lett. 59, 305-309
Beier, H.M. (1973) Die hormonelle Steuerung der Uterussekre-
tion und frühen Embryonalentwicklung des Kaninchens.
Habilitationsschrift, Medical Faculty, Kiel
Beier, H.M. (1974) Oviducal and uterine fluids.
J.Reprod.Fert. 37, 221-237
Beier, H.M. (1976) Uteroglobin and related biochemical changes
in the reproductive tract during early pregnancy in the
rabbit.
J.Reprod.Fert.(Suppl.) 25, 53-63
Beier, H.M. (1978) Physiology of uteroglobin. In "Novel aspects
of reproductive physiology" (Ed. C.H. Spilman & J.W. Wilks)
(Spectrum Publ.Inc., New York,London) pp. 220-248
Beier, H.M. (1979) Die Entwicklung der Blastozyste: Differen-
zierung von Embryoblast und Trophoblast.
Verh.Anat.Ges. 73, 353-378
Beier, H.M. & Maurer, R.R. (1975) Uteroglobin and other proteins in rabbit blastocyst fluid after development in vivo
and in vitro.
Cell Tiss.Res. 159, 1-10
Beier, H.M., Mootz, U., Fischer, B. & Ströbele-Müller, R.
(1983) Growth and differentiation of rabbit blastocysts in
defined culture media. In "Fertilization of the human egg

in vitro" (Ed. H.M. Beier & H.R. Lindner) (Springer-Verlag,
Berlin, Heidelberg, New York, Tokyo) pp. 371-386
Beier, H.M., Mootz, U. & Kühnel, W. (1972) Asynchrone Eitrans-
plantationen während der verzögerten Uterussekretion beim
Kaninchen.
Proc. VII.Int.Congr.Anim.Reprod.Art.Insem., München, Vol. 3,
pp. 1891-1896
Boué, J., Boué, A. & Lazar, P. (1975) The epidemiology of human
spontaneous abortions with chromosomal anomalies. In "Aging
gametes" Int.Symp.Seattle 1973 (Karger, Basel) pp. 330-348
Bullock, D.W., Woo, S.L.C. & O'Malley, B.W. (1976) Uteroglobin
messenger RNA. Translation in vitro.
Biol.Reprod. 15, 435-443
Chang, M.C. (1950) Development and fate of transferred rabbit
ova or blastocysts in relation to ovulation time of recipients.
J.exp.Zool. 114, 197-216
Cook, B. & Hunter, R.H.F. (1978) Systemic and local hormonal
requirements for implantation in domestic animals.
J.Reprod.Fert. 54, 471-482
Daniel, J.C.,Jr. (1971) Growth of the preimplantation embryo
of the northern fur seal and its correlation with changes
in uterine proteins.
Devl.Biol. 26, 316-331
Delbos, R., Fischer, B. & Beier, H.M. (1981) Influence of
superovulation on uterine secretion protein patterns in
pseudopregnant rabbits.
Proc. III. World Congr. Human Reprod., Berlin, p. 229
Denker, H.-W. (1977) Implantation: The role of proteinases,
and blockage of implantation by proteinase inhibitors.
(Advances in anatomy, embryology and cell biology, Vol. 53/5)
(Springer-Verlag, Berlin, Heidelberg, New York)
Denker, H.-W. (1982) Proteases of the blastocyst and of the
uterus. In "Proteins and steroids in early pregnancy"
(Ed. H.M. Beier & P. Karlson) (Springer-Verlag, Berlin,
Heidelberg, New York) pp. 183-208
Dixon, S.N. & Gibbons, R.A. (1979) Proteins in the uterine
secretions of the cow.
J.Reprod.Fert. 56, 119-127
El-Banna, A.A. & Daniel J.C.,Jr. (1972a) Stimulation of rabbit
blastocysts in vitro by progesterone and uterine proteins
in combination.
Fert.Steril. 23, 101-104
El-Banna, A.A. & Daniel, J.C.,Jr. (1972b) The effect of pro-
tein fractions from rabbit uterine fluids on embryo growth
and uptake of nucleic acid and protein precursors.
Fert.Steril. 23, 105-114
Enders, R.K. & Enders, A.C. (1963) Morphology of the female
reproductive tract during delayed implantation in the mink.
In "Delayed implantation" (Ed. A.C. Enders) (University of
Chicago Press, Chicago) pp. 129-139
Fazleabas, A.T., Bazer, F.W. & Roberts, R.M. (1982) Purifica-
tion and properties of a progesterone-induced plasmin/tryp-
sin inhibitor from uterine secretion of pigs and its immu-
nocytochemical localization in the pregnant uterus.
J.biol.Chem. 257, 6886-6897

106

Fischer, B., Denker, H.-W. & Beier, H.M. (1982) Incomplete
transformation of rabbit embryo coverings in vitro.
Proc.Ann.Conf.Soc.Study Fertility, Sutton Bonnington, 58,
abstr. 93
Fischer, B., Denker, H.-W., Beier, H.M. & Gerdes, H.J. (1983)
Der Umbau der extrazellulären Blastozystenhüllen als ein
empfindlicher Indikator für unphysiologische Milieubedin-
gungen in der In-vitro-Kultur.
Verh.Anat.Ges. 77, 427-429
Fischer, B., Mootz, U. & Beier, H.M. (1981) Morphology of
rabbit embryos recovered from ligated oviducts.
Proc. III. World Congr.Human Reprod., Berlin, p. 217
Fishel, S.B. (1980) Radiolabelled uterine proteins during
early pregnancy and pseudopregnancy in mice after unila-
teral ovariectomy and superovulation.
J.Reprod.Fert. 59, 473-478
Garcia, J.F. (1983) Implantation failure after embryo transfer
in in vitro fertilization.
Infertility 6, 187-199
Godkin, J.D., Bazer, F.W. & Roberts, R.M. (1984) Ovine tropho-
blast protein 1, an early secreted blastocyst protein,
binds specifically to uterine endometrium and affects pro-
tein synthesis.
Endocrinology 114, 120-130
Gordon, J. (1982) Synchronization of estrus and superovulation
in cattle. In "Mammalian egg transfer" (Ed. C.E. Adams)
(CRC Press Inc., Boca Raton) pp. 63-80
Heap, R.B. (1980) Implantation. In "Steroid induced uterine
proteins" (Ed. M. Beato) (Elsevier/North-Holland Biomedical
Press, Amsterdam, New York, Oxford) pp. 35-46
Heap, R.B., Flint, A.P.F., Gadsby, J.E. & Rice, C. (1979)
Hormones, the early embryo and the uterine environment.
J.Reprod.Fert. 55, 267-275
Heap, R.B., Gadsby, J.E., Rice, C. & Perry, J.S. (1982)
The synthesis of steroids and proteins in the pig blasto-
cyst. In "Proteins and steroids in early pregnancy"
(Ed. H.M. Beier & P. Karlson) (Springer-Verlag, Berlin,
Heidelberg, New York) pp. 157-171
Kirby, D.R.S. (1962) The influence of the uterine environment
on the development of mouse eggs.
J.Embryol.exp.Morph. 10, 496-506
Kirby, D.R.S. (1969) The extra-uterine mouse egg as an experi-
mental model.
Adv.Biosci. 4, 255-273
Kirchner, C., Hirschhäuser, C. & Kionke, M. (1971) Protease
activity in rabbit uterine secretion 24 hours before im-
plantation.
J.Reprod.Fert. 27, 259-260
Laster, D.B. (1977) A pregnancy-specific protein in the bovine
uterus.
Biol.Reprod. 16, 682-690
Lauritsen, J.G. (1982) The cytogenetics of spontaneous abortion
Res.Reprod. 14, 3-4
Leeton, J., Trounson, A., Wood, C. & Gianaroli, L. (1983)
In vitro fertilization and embryo-transfer: the Monash group
experiences 1981-83.
Acta Eur.Fert. 14, 95-100

Martin, L. (1984) On the source of uterine 'luminal fluid' pro-
teins in the mouse.
J.Reprod.Fert. 71, 73-80
Maurer, R.R. & Beier, M.M. (1976) Uterine proteins and develop-
ment in vitro of rabbit preimplantation embryos.
J.Reprod.Fert. 48, 33-41
Milligan, S.R. & Martin, L. (1984) The resistance of the mouse
uterine lumen to flushing and possible contamination of
samples by plasma and interstitial fluid.
J.Reprod.Fert. 71, 81-87
Moore, N.W. (1982) Egg transfer in the sheep and goat.
In "Mammalian egg transfer" (Ed. C.E. Adams) (CRC Press
Inc., Boca Raton) pp.119-133
Mulholland, J. & Villee, C.A.,Jr. (1984) Proteins synthesized
by the rat endometrium during early pregnancy.
J.Reprod.Fert. 72, 395-400
Murdoch, B.E. & O'Shea, T. (1977) ß-Glucuronidase in the re-
productive tract of the ewe during early pregnancy.
Theriogenology 8, 140
Murdoch, B.E. & O'Shea, T. (1978) Activity of enzymes in the
mucosal tissues and rinsings of the reproductive tract of
the naturally cyclic ewe.
Aust. J.biol.Sci. 31, 345-354
Murdoch, R.N. (1970) Uterine endometrial phosphomonoesterases
in relation to implantation in the ewe and rabbit doe.
Aust. J.biol.Sci. 23, 1089-1097
Murdoch, R.N., Kay, D.J. & Cross, M. (1978) Activity and sub-
cellular distribution of mouse uterine alkaline phosphatase
during pregnancy and pseudopregnancy.
J.Reprod.Fert. 54, 293-300
Murdoch, R.N. & White, I.G. (1968) Activities of enzymes in
the endometrium, caruncles and uterine rinsings of pro-
gesteron-treated and naturally cycling ewes.
Aust. J.biol.Sci. 21, 123
Murray, F.A.,Jr., Bazer, F.W., Rundell, J.W., Vincent, C.K.,
Wallace, H.D. & Warnick, A.C. (1971) Developmental failure
of swine embryos restricted to the oviducal environment.
J.Reprod.Fert. 24, 445-448
Peplow, P.V. (1982) Analysis of endopeptidase and arylamidase
enzymes in uterine fluid of oestrogen-treated rats.
J.Reprod.Fert. 66, 649-654
Ponstingl, H., Nieto, A. & Beato, M. (1978) Amino acid se-
quence of progesterone-induced rabbit uteroglobin.
Biochemistry 17, 3908-3912
Pratt, H.P.M. (1977) Uterine proteins and the activation of
embryos from mice during delayed implantation.
J.Reprod.Fert. 50, 1-8
Roberts, G.P. & Parker, J.M. (1974a) Macromolecular components
of the luminal fluid from the bovine uterus.
J.Reprod.Fert. 40, 291-303
Roberts, G.P. & Parker, J.M. (1974b) An investigation of en-
zymes and hormone-binding proteins in the luminal fluid
of the bovine uterus.
J.Reprod.Fert. 40, 305-313

Roberts, G.P., Parker, J.M. & Symonds, H.W. (1976) Macromolecular components of genital tract fluids from the sheep.
J.Reprod.Fert. 48, 99-107

Roberts, R.M. & Bazer, F.W. (1980) The properties, function
and hormonal control of synthesis of uteroferrin, the
purple protein of the pig uterus. In "Steroid induced
uterine proteins" (Ed. M. Beato) (Elsevier/North-Holland
Biomedical Press, Amsterdam, New York, Oxford) pp. 35-46

Rowson, L.E.A. (1971) The role of reproductive research in
animal production.
J.Reprod.Fert. 26, 113-126

Rowson, L.E.A. & Moor, R.M. (1966) Embryo transfer in the
sheep: the significance of synchronising oestrus in the
donor and recipient animal.
J.Reprod.Fert. 11, 207-212

Rowson, L.E.A., Moor, R.M. & Lawson, R.A.S. (1969) Fertility
following egg transfer in the cow: effect of method, medium,
and synchronization of oestrus.
J.Reprod.Fert. 18, 517-523

Squire, G.D., Bazer, F.W. & Murray, F.A.,Jr. (1972) Electrophoretic patterns of porcine uterine protein secretions
during the estrous cycle.
Biol.Reprod. 7, 321-325

Ströbele-Müller, R. (1984) Die Blastocysten-Entwicklung in der
in-vitro-Kultur unter dem Einfluß von fötalen Kälberserum-
proteinen.
MD-Thesis, Medical Faculty, RWTH Aachen

Strzemienski, P.J., Do, D. & Kenney, R.M. (1984) Antibacterial
activity of mare uterine fluid.
Biol.Reprod. 31, 303-311

Surani, M.A.H. (1976) Uterine luminal proteins at the time of
implantation in rats.
J.Reprod.Fert. 48, 141-145

Surani, M.A.H. (1977) Cellular and molecular approaches to
blastocyst uterine interactions at implantation.
In "Development in mammals" (Ed. M.H. Johnson) (North-
Holland Publish.Comp., Amsterdam, New York, Oxford) Vol. 1,
pp. 245-305

Zavy, M.T., Clark, W.R., Sharp, D.C., Roberts, R.M. & Bazer,
F.W. (1982a) Comparison of glucose, fructose, ascorbic acid
and glucosephosphate isomerase enzymatic activity in uterine flushings from nonpregnant and pregnant gilts and pony
mares.
Biol.Reprod. 27, 1147-1158

Zavy, M.T., Roberts, R.M. & Bazer, F.W. (1984) Acid phosphatase and leucine aminopeptidase activity in the uterine
flushings of non-pregnant and pregnant gilts.
J.Reprod.Fert. 72, 503-507

Zavy, M.T., Sharp, D.C., Bazer, F.W., Fazleabas, A., Sessions,
F. & Roberts, R.M. (1982b) Identification of stage-specific
and hormonally induced polypeptides in the uterine protein
secretions of the mare during the oestrus cycle and pregnancy.
J.Reprod.Fert. 64, 199-207

ACCUMULATION OF SEX STEROIDS IN PREIMPLANTATION PIG BLASTOCYSTS AND THEIR IMPORTANCE FOR AN UNDISTURBED EARLY EMBRYONIC DEVELOPMENT

H. Niemann and F. Elsaesser

Institut für Tierzucht und Tierverhalten (FAL), Mariensee,
3057 Neustadt 1, F.R.G.

ABSTRACT

The accumulation of ^{3}H-progesterone and ^{3}H-estradiol-17ß in five day-old pig embryos was studied in vitro. After 6 hrs. of incubation, the uptake of ^{3}H-progesterone was 132 ± 18 cpm ($\bar{x} \pm$ SEM) and apparently was nonspecific since no significant inhibition could be determined in the presence of a 100-fold excess of unlabelled progesterone or estradiol-17ß. After 6 hrs. of incubation, the uptake of ^{3}H-estradiol-17ß was 134 ± 25 cpm, but apparently was specific since a significant ($p < 0.01$) competitive inhibition (67.7 %) could be determined in the presence of a 100-fold excess of unlabelled estradiol-17ß. Apparent specific binding was calculated to be 0.87 fmoles estradiol-17ß per 10 blastocysts or 72.5 fmoles estradiol-17ß per mg protein. After 2 hrs. of incubation, a significant ($p < 0.001$) competitive inhibition (58.4 %) of ^{3}H-estradiol-17ß uptake was present, but not after 20 min. of culture. The uptake of ^{3}H-estradiol-17ß was significantly ($p < 0.001$) lower (52 ± 5 cpm) in degenerate embryos and unfertilized ova from day 5 than by viable blastocysts from the same day. In addition, competitive inhibition was lacking. The results suggest the presence of specific binding sites for estradiol-17ß in the early pig embryo which are still unidentified. In-vitro development of pig morulae was inhibited by adding the anti-estrogen Nafoxidine to the culture medium. The inhibition could be overcome by adding estradiol-17ß to the medium. Estradiol-17ß appears to have a distinct physiological function for an undisturbed development of the early pig embryo, a suggestion which is going to be substantiated in subsequent experiments.

INTRODUCTION

It is well known, that progesterone and estradiol-17ß are important for the early embryonic development, in particular for a physiological transport of fertilized ova through the oviduct (for review see Hunter, 1980). However, it is not clear whether these hormones affect the embryos directly or whether their action is on the oviduct or uterus which then secrete substances that act on the embryos. Direct effects are conceivable because both hormones have been found in preimplantation embryos from mice and rabbits (Seamark and Lutwak-Mann,

1972; Borland et al., 1977; Grube et al., 1978; Singh and Booth, 1978; Angle and Mead, 1979). To investigate the mode of action and the physiological function of both steroids inside the embryo, we studied the accumulation of both progesterone and estradiol-17ß and the effects of estradiol withdrawal in the early pig blastocyst.

MATERIALS AND METHODS

Morula stage embryos were collected at slaughter from superovulated German Landrace gilts on day 5 after insemination. Embryos evaluated as viable were cultured in vitro for 24 hrs. up to the blastocyst stage (Niemann et al., 1983). The embryos were then incubated for 20, 120 or 360 minutes in medium containing 400.000 or 800.000 cpm (= counts per minute) $(1, 2, 6, 7 - {}^3H)$ progesterone ($\left[{}^3H-\right] P_4$) or $(2, 4, 6, 7 - {}^3H)$ estradiol-17ß ($\left[{}^3H-\right] E_2$). Uptake specificity was tested in parallel by determining the uptake in presence of a 100-fold excess (1 µM) of unlabelled progesterone ($\left[{}^3H-\right] P_4+P_4$; $\left[{}^3H-\right]E_2+P_4$) or estradiol-17ß ($\left[{}^3H-\right] E_2+E_2$; $\left[{}^3H-\right] P_4+E_2$). At termination of incubation embryos were washed five times in Krebs-Ringer bicarbonate (KRB) and were then transferred into vials containing tissue solvent. Radioactivity in embryos was counted in a liquid scintillation spectrometer. Total activity in the incubation medium and background activity of the last washing medium was always determined. The latter was subtracted from the activity found in the embryos. Competitive inhibition was calculated as the difference between total uptake of ${}^3H$-progesterone or ${}^3H$-estradiol-17ß and uptake in the presence of a 100-fold excess of unlabelled progesterone or estradiol-17ß. A more detailed description of the methods is given by Niemann and Elsaesser (1984).

UPTAKE AND SPECIFICITY OF ${}^3H$-PROGESTERONE

After 6 hrs. in-vitro incubation, uptake of ${}^3H$-progesterone amounted to 132 $\pm$ 18 cpm per 10 blastocysts (Fig. 1). Addition of a 100-fold excess of unlabelled progesterone or estradiol-17ß to the incubation medium resulted in a slight and nonsignificant reduction of the uptake of ${}^3H$-progesterone. The competitive inhibition was calculated to be 27.0 % and 20.1 % for progesterone and estradiol-17ß respectively. Thus, the uptake of progesterone by the pig blastocyst apparently is nonspecific only.

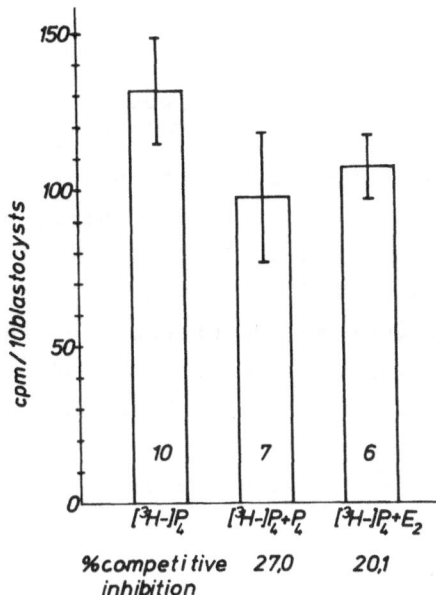

Fig. 1 Uptake and its specificity ($\bar{x} \pm$ SEM) of ^{3}H-progesterone by pig blastocysts after 6 hrs. of incubation. The number of experiments, each involving 5-10 embryos, is indicated at the bottom of the bars (data from Niemann and Elsaesser, 1984).

Similar to our results, other investigators (Angle and Mead, 1979) observed a nonspecific uptake of progesterone by rabbit embryos, which primarily took place into the blastocoelic fluid with maximum concentrations after 2 hrs. Progesterone that accumulates in rabbit blastocysts in the uterus probably originates from the uterine fluid (Borland et al., 1977). It may well be that progesterone uptake also occurs nonspecifically into the blastocoelic fluid of the pig blastocyst. Progesterone which is necessary as a precursor of estradiol and estrone synthesis may have importance for the pig embryo from days 10 to 12, since estrogens are needed as an embryonic signal for the maternal recognition of pregnancy (Heap et al., 1979). But so far, no conclusive statements are possible regarding the kind of specific function progesterone has for the very early embryonic stages in the pig.

UPTAKE AND SPECIFICITY OF ^{3}H-ESTRADIOL-17β

Estrogens obviously are necessary for implantation to occur, since implantation response was significantly higher when mice blastocysts treated with 10 nM estradiol-17β in-vitro were transferred to foster mothers than when untreated blastocysts were transferred (Smith, 1968). Estrogens, detected in blastocysts were considered to be of maternal or embryonic origin (Borland et al., 1977; Singh and Booth, 1978; Dickmann et al., 1976). Estrogens may be produced by the embryo itself, since the enzymatic capacity for the synthesis of estrogens has been demonstrated in preimplantation embryos from rodents and farm animals (Dickmann et al., 1976; Hoversland et al., 1982; Perry et al., 1973; Gadsby et al., 1976).

The present experiments show, that similar to progesterone the uptake of ^{3}H-estradiol-17β exceeded 130 cpm per 10 blastocysts after 6 hrs. of incubation (Fig. 2). But in contrast to progesterone, addition of a 100-fold excess of unlabelled estradiol-17β reduced the uptake of radiolabelled estradiol-17β significantly by 67.7 %. In the presence of a 100-fold excess of unlabelled progesterone a slight and nonsignificant inhibition (23.3 %) of the uptake of ^{3}H-estradiol-17β could be determined. Based on a specific activity of 93.2 Ci/mM estradiol-17β total and specific uptake by 10 blastocysts were calculated to be 1.31 fmoles and 0.87 fmoles respectively of estradiol-17β.

Further experiments with ^{3}H-estradiol-17β revealed that after 2 hrs. of incubation the uptake was only nonsignificantly less (82 $\pm$ 2 cpm) than after 6 hrs. of incubation and the significant (p < 0.001) competitive inhibition (58.4 %) was still maintained (Fig. 3). In contrast, after 20 minutes of incubation a significantly (p < 0.005) reduced uptake of 41 $\pm$ 4 cpm was found (Fig. 4) when compared to an incubation period of 2 or 6 hrs. and no competitive inhibition could be determined. The uptake of ^{3}H-estradiol-17β was also significantly

(p < 0.001) less (52 $\pm$ 5 cpm) in degenerate embryos and unfertilized ova recovered on day 5 when compared to viable blastocysts after 2 or 6 hrs. of incubation. No competitive inhibition was apparent (Fig. 5).

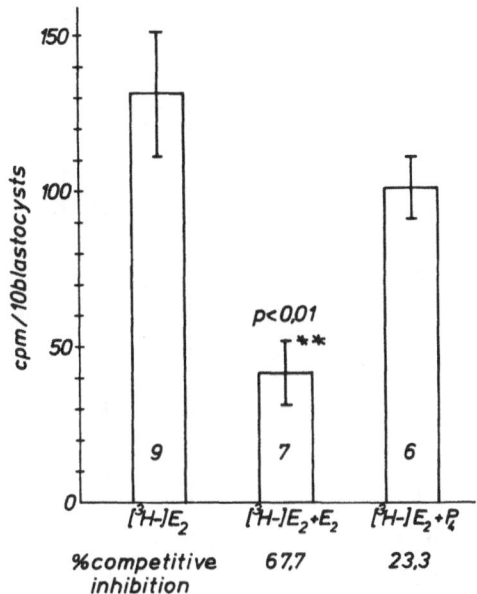

Fig. 2 Uptake and its specificity ($\bar{x}$ $\pm$ SEM) of ^{3}H-
estradiol-17β by pig blastocysts after 6 hrs. of incubation. The number of experiments, each involving 10-20
embryos, is indicated at the bottom of the bars (data
from Niemann and Elsaesser, 1984).

Furthermore, the uptake of ^{3}H-estradiol-17β was tested
after 6 hrs. of culture in a protein-free culture medium. The
amount was similar to those embryos incubated in complete medium and the competitive inhibition was maintained (Fig. 6).

In contrast to the accumulation of progesterone, the present results indicate an apparent specific uptake and binding
of estradiol-17β by the pig blastocyst. The finding of a specific estradiol-17β uptake is supported by the significant competitive inhibition in presence of a 100-fold excess of unlabelled estradiol-17β and furthermore when binding is calculated on a protein basis. Assuming that 1.2 µg of protein are
present in a pig blastocyst (Wright et al., 1983) then approxi-

mately 72.5 fmoles of estradiol-17β are bound per mg protein.
This is in the order of magnitude described for instance for
the binding capacity of rat pituitary cytosol for estradiol-
17β (Van Beurden-Lamers et al., 1974).

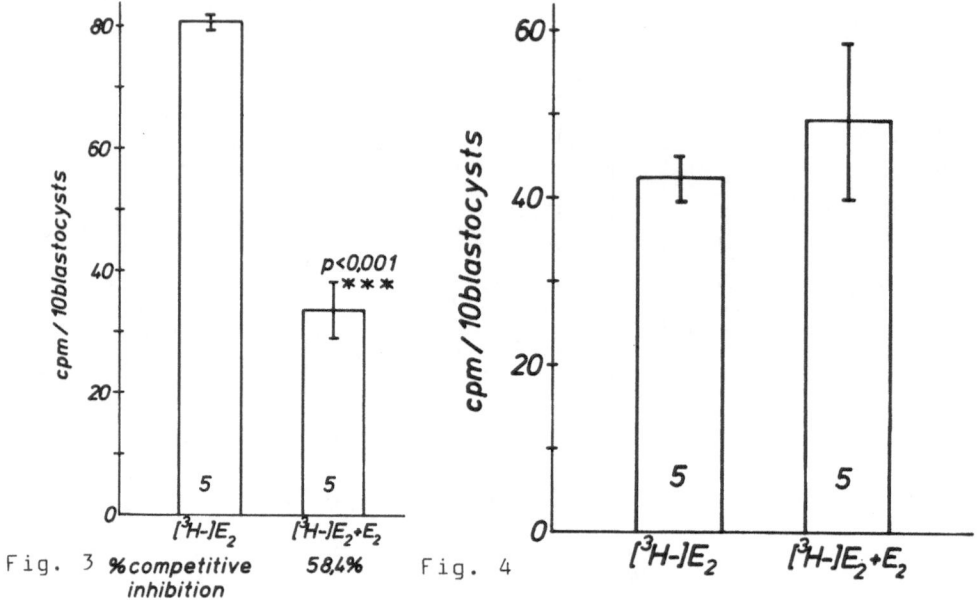

Fig. 3 %competitive 58.4% Fig. 4
inhibition

Fig. 3 Uptake and its specificity (x̄ ± SEM) of ³H-
estradiol-17β by pig blastocysts after 2 hrs. of incubation. The number of experiments, each involving
10 embryos, is indicated at the bottom of the bars
(data from Niemann and Elsaesser, 1984).

Fig. 4 Uptake and its specificity (x̄ ± SEM) of ³H-
estradiol-17β by pig blastocysts after 20 min. of incubation. The number of experiments, each involving
10-20 embryos, is indicated at the bottom of the bars
(data from Niemann and Elsaesser, 1984).

Holmes (1976) reported specific binding of estradiol-17β
in cytosol of rabbit blastocysts. Bhatt and Bullock (1974)
also suggested the presence of a specific binding protein for
estradiol-17β in rabbit blastocysts, although they did not
prove the existence of a receptor protein. Smith (1968) reported an uptake of ³H-estradiol-17β by mouse blastocysts without
any details on specificity. Other investigators found no sa-

turation or competitive inhibition and suggested a nonspecific uptake in mouse embryos (Habenicht, 1982). In the hamster, an increasing uptake of radiolabelled estradiol-17ß from the 4-cell to the blastocyst-stage embryo was observed within 2 hrs. of in-vitro culture. A significantly smaller uptake in embryos recovered from superovulated donors was explained by a satura-tion of receptors (Hutz and Dukelow, 1982).

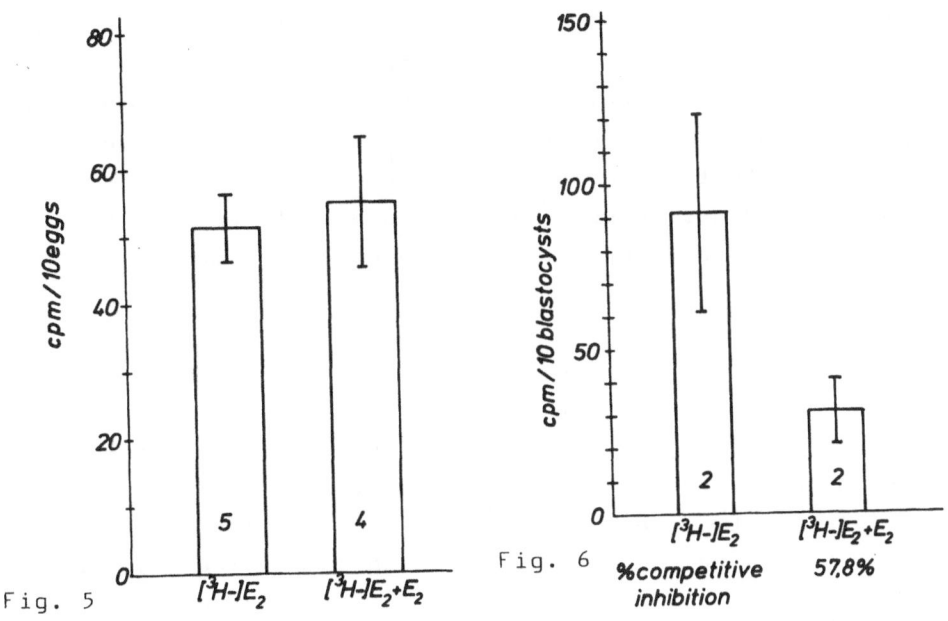

Fig. 5

Fig. 6

Fig. 5 Uptake and its specificity ($\bar{x} \pm$ SEM) of ^{3}H-estradiol-17ß by unfertilized ova and degenerate em-bryos after 6 hrs. of incubation. The number of expe-riments, each involving 20 embryos, is indicated at the bottom of the bars (data from Niemann and Elsaesser, 1984).

Fig. 6 Uptake and its specificity ($\bar{x} \pm$ SEM) of ^{3}H-estradiol-17ß by pig blastocysts after 6 hrs. of in-cubation in protein-free culture medium. The number of experiments, each involving 20 embryos, is indi-cated at the bottom of the bars (data from Niemann and Elsaesser, 1984).

According to our results, estradiol-17ß apparently is bound to specific binding sites in the early pig embryo which so far are still unidentified. Uptake and specificity of estra-

diol-17β appear to be dependent on the viability of the blastocyst, suggesting a blastocyst controlled mechanism. The amount
and specificity of uptake of estradiol-17β seem to be independent of proteins present in the surrounding medium, since uptake and inhibition were still present in protein-free culture
medium. Preliminary unpublished results suggest that estrone
is taken up by the pig blastocyst in a saturable manner as
well.

Estradiol-17β, apart from its role as an embryonic signal
for the maternal recognition of pregnancy in the pig may have
a specific function at even earlier stages of development. It
is suggested that estradiol-17β is necessary for certain functional changes during cell division and differentiation in particular for the development from the compacted morula to the
blastocyst stage.

EFFECTS OF THE ANTI-ESTROGEN NAFOXIDINE ON IN-VITRO DEVELOPMENT OF PIG MORULAE (preliminary results)

To substantiate this suggestion the effects of the anti-
estrogen Nafoxidine (nonsteroidal, 1-2, p-3, 4 dihydro-methoxy-
2-phenyl-1-naphthyl-phenoxyethyl) upon the development of pig
morulae were studied in-vitro (Niemann and Elsaesser, unpublished observation). Recently Nafoxidine has been shown to inhibit in-vitro blastocyst expansion in the rat (Roy et al.,
1982). In our experiment a culture medium without BSA but supplemented with 10 % heat-inactivated lamb serum was used. Whereas
control embryos developed undisturbed (95 % grew to blastocysts),
addition of 15 µg/ml Nafoxidine to the culture medium resulted
in an almost complete inhibition of the embryonic development,
with only 10 % of the morula stages developing to blastocysts.
Lower concentrations of Nafoxidine obviously did not affect
embryonic development, increasing Nafoxidine concentrations led
to a complete inhibition of the development. However, the inhibitory effect could be overcome by adding estradiol-17β to
the medium. The percentage of development increased to around
60 %. This effect obviously was specific for estradiol-17β,
since the inhibition could not be overcome by adding pro-

gesterone or cortisol to the medium.

These data support the suggestion of a specific role of estradiol-17ß in the development of the early pig embryo. Obviously estradiol-17ß is needed for an undisturbed development from the morula to the blastocyst stage. Studies are in progress to substantiate this intriguing possibility.

REFERENCES

Angle, M.J. and Mead, R.A. 1979. The source of progesterone in preimplantation rabbit blastocysts. Steroids 33, 625-637.
Bhatt, B.M. and Bullock, D.W. 1974. Binding of oestradiol to rabbit blastocysts and its possible role in implantation. J. Reprod. Fert. 39, 65-70.
Borland, R.M., Erickson, G.F. and Ducibella, T. 1977. Accumulation of steroids in rabbit preimplantation blastocysts. J. Reprod. Fert. 49, 219-224.
Dickmann, Z., Dey, S.K. and Sengupta, J. 1976. A new concept: control of early pregnancy by steroid hormones originating in the preimplantation embryo. Vitams Horm. 34, 215-242.
Gadsby, J.E., Burton, R.D., Heap, R.B. and Perry, J.S. 1976. Steroid metabolism and synthesis in early embryonic tissue of pig, sheep and cow. J. Endocr. 71, 45-46.
Grube, K.E., Gwazdauskas, F.C., Lineweaver, J.A. and Vinson, W.E. 1978. Steroidogenic capabilities of the early mouse embryo. Steroids 32, 345-354.
Habenicht, U.-F. 1982. The binding of oestrogens to mouse blastocysts. Acta Endocrinologica 99, (Suppl. 246) 21 (Abstr.).
Heap, R.B., Flint, A.P.F., Gadsby, J.E. and Rice, C. 1979. Hormones, the early embryo and the uterine environment. J. Reprod. Fert. 55, 267-275.
Holmes, P.V. 1976. Evidence of cytosol receptor protein specific for estradiol in the preimplantation blastocyst. VIII. Intern. Congress on Anim. Reprod. and A.I., Cracow 3, 290-293.
Hoversland, R.C., Dey, S.K. and Johnson, D.C. 1982. Aromatase activity in the rabbit blastocyst. J. Reprod. Fert. 66, 259-263.
Hunter, R.H.F. 1980. Function of the fallopian tubes in relation to gametes and embryos. In "Physiology and Technology of reproduction in female domestic animals" (Ed. R.H.F. Hunter). (Academic Press, London). pp. 145-183.
Hutz, R.J. and Dukelow, W.R. 1982. Macromolecular synthesis and steroid uptake during early embryonic development in the hamster and squirrel monkey. Proc. Soc. Study Reprod., Madison (Wisconsin), pp. 66A (Abstr.).
Niemann, H., Illera, M.J. and Dziuk, P.J. 1983. Developmental capacity, size and number of nuclei in pig embryos cultured in vitro. Anim. Reprod. Sci. 5, 311-321.

Niemann, H. and Elsaesser, F. 1984. Uptake and effects of ova-
    rian steroids in the early pig embryo: in vitro and in
    vivo studies. Theriogenology 21, 84-102.
Perry, J.S., Heap, R.B. and Amoroso, R.C. 1973. Steroid hor-
    mone production by pig blastocysts.Nature (Lond.) 245,
    45-47.
Roy, S.K., Sengupta, J., Paria, B.C. and Manchanda, S.K. 1982.
    In vitro inhibition of trophoblast maturation and expan-
    sion of early rat blastocysts by an oestrogen antagonist.
    Acta Endocrinologica 99, 129-135.
Seamark, R.F. and Lutwak-Mann, C. 1972. Progestins in rabbit
    blastocysts. J. Reprod. Fert. 29, 147-148.
Singh, M.M. and Booth, W.D. 1978. Studies on the metabolism of
    neutral steroids by preimplantation rabbit blastocysts in
    vitro and the origin of blastocyst oestrogen. J. Reprod.
    Fert. 53, 297-304.
Smith, D.M. 1968. The effect on implantation of treating cul-
    tured mouse blastocysts with oestrogen in vitro and the
    uptake of $[^3H]$ oestradiol by blastocysts. J. Endocr.
    41, 17-29.
Van Beurden-Lamers, W.M.O., Brinkmann, A.O., Mulder, E. and
    Van der Molen, H.J. 1974. High affinity binding of oestra-
    diol-17β by cytosol from testis interstitial tissue, pi-
    tuitary adrenal, liver and accessory sex glands of the
    male rat. Biochem. J. 140, 495-502.
Wright, R.W., Jr., Grammer, J., Bondioli, K., Kuzan, F. and
    Menino, A., Jr. 1983. Protein content and volume of early
    porcine blastocysts. Anim. Reprod. Sci. 5, 207-212.

PERIOVULATORY HORMONE PROFILES IN RELATION TO
EMBRYONIC DEVELOPMENT AND MORTALITY IN PIGS

F. Helmond, A. Aarnink, C. Oudenaarden

Department of Animal Physiology
Agricultural University 6709 PJ Haarweg 10
Wageningen, The Netherlands

ABSTRACT

The timing between the onset of oestrus and the preovulatory Luteinizing hormone (LH) peak in gilts was investigated. The LH peak was most frequently observed approximately 10h after the onset of oestrus, although large variations from 20h before to more than 24h after the onset of oestrus were noticed. The time interval between the LH peak and ovulation, however, was rather constant and so a large variation with regard to the time interval between the onset of oestrus and ovulation was observed. Insemination of pigs at a fixed time interval after the onset of oestrus resulted in conceptions with sperm of variable age. The results suggest a higher incidence of embryonic mortality with an increasing interval between the time of insemination and ovulation. The rate of development of pig embryos between day 3 and 7 of pregnancy was studied with respect to the time of ovulation. The results indicate that pig embryos double their number of cells every 11 hours.

INTRODUCTION

The moment of artificial insemination in swine is generally timed with respect to the onset of oestrus. This assumes a fixed time interval between the onset of oestrus and the time of the LH peak since the interval between the LH peak and ovulation seems to be constant (Liptrap and Raeside 1966; Hunter 1967). Niswender et al., (1970) reported that the LH peak coincides with the onset of oestrus whereas Tilton et al., (1982), reported that the LH peak varies between 2h before to 22 h after the onset of oestrus and Ziecik et al., (1982) reported that the LH peak occurs in 7 out of 14 sows between 8 and 32 h before the onset of oestrus. The LH peak may thus vary between 32h before to 22h after the onset of oestrus. Artificial insemination at a fixed time interval after the onset of oestrus may thus result in a conception with "aged" sperm or ova due to a "late" or "early" LH peak with respect to the onset of oestrus respectively. The aim of this study therefore is firstly, to reinvestigate the time relationship between the LH peak and onset of oestrus and secondly, to investigate the consequences of the variability between the time of the LH peak and the onset of oestrus with regard to both fertilization rate and embryonic development and mortality.

MATERIALS AND METHODS

Thirty-three oestrous cycles in 27 gilts of the Dutch Landrace type were studied. The animals were cannulated via the cephalic vein some days before the expected onset of oestrus. Bloodsamples were drawn once daily except in the period between 2 days before to 3 days after the onset of oestrus. In this period the frequency of bloodsampling was increased to every 6 hours. Oestrus was checked with a vasectomized boar in the same frequency as the sampling of the blood.

The onset of oestrus is defined as the point between the time when mating was first observed and the time of the preceeding oestrus check. The animals were inseminated between 8 and 32h after the onset of oestrus. Embryos and corpora lutea were collected between day 3 and 10 of pregnancy. The time of ovulation is calculated from the time that progesterone level increases to 1 ng/ml above its basal level.

Bloodsamples were assayed in duplicate for LH, progesterone and oestradiol using specific radioimmunoassays. Oestradiol and progesterone were determined in virtually the same way as described before (Helmond et al., 1980). Concentrations of LH were estimated in duplicate in 200 µl plasma by a homologous double-antibody radioimmunoassay using porcine LER 786-3 (potency 0.65xNIH-LH-S1) as a standard and for iodination. Anti-porcine LH (GND 566) was used at a 1:32,000 working dilution. The sensitivity of the assay was 0.5 ng/ml at the 90% B/B_o level.

RESULTS

The distribution of the time interval between the LH peak and the onset of oestrus is shown in Fig. 1.

Fig. 1. The onset of oestrus in relation to the LH peak in pigs.

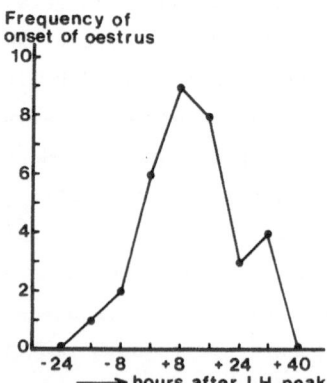

Onset of oestrus in relation to the LH peak in pigs (N=33).

The LH peak occurred on average 10±13h (SD) after the onset of oestrus with a range of 20h before to 32.5h after the onset of oestrus. The interval between the LH peak and ovulation was 42±5h (SD) with a range of 31-53h. The interval between oestrus and ovulation was 55±13h (SD) with a range of 18-76h.

Twenty-seven gilts were inseminated in the period between 52h before to 20h after ovulation. Eighteen of these animals became pregnant. In the remaining 9 animals the fertilization failure was probably due to "overaged" sperm (3 gilts) or "overaged ova" (4 gilts); compare Fig. 2A and 2B.

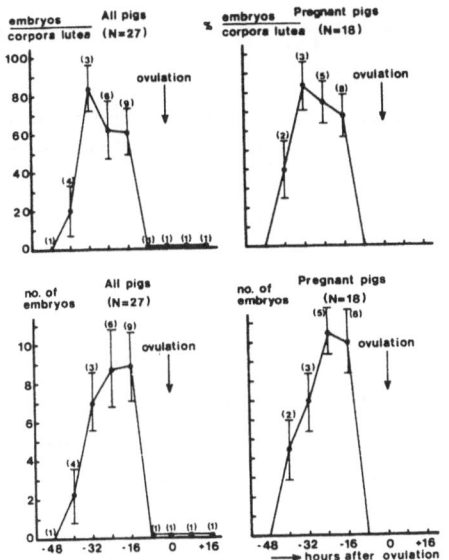

Fig. 2. The interval between insemination and ovulation in relation to embryonic mortality in pigs.

In the pregnant animals a sharp decline in embryonic survival was noticed when the interval between insemination and ovulation increased to over 36h (Fig. 2B). The average number of embryos declined sharply after an interval of over 28h (see Fig. 2D).

The rate of development of the pig embryos was calculated by plotting the natural logarithm (ln) of the number of cells (y-value) against their age in hours after ovulation (x-value). The rate of development between 60 and 130h after ovulation can be described after linear regression analyses by y=0.062x-2.34 (n=67. r=0.902, p<0.001). From Fig. 3 it may be concluded that a pig embryo doubles its number of cells once every hour between 60 and 130h after ovulation. Before 60h after ovulation the development of the pig embryo is arrested in the so-called "4-cell block".

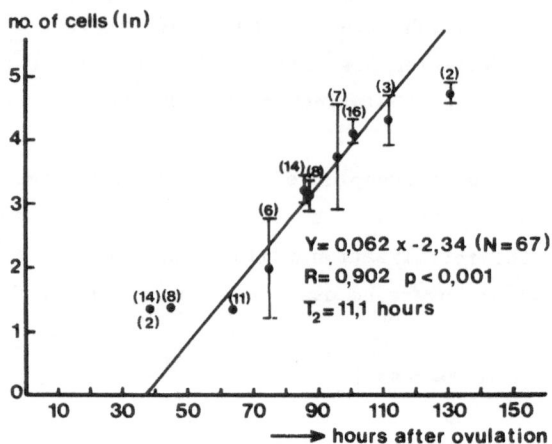

Fig.3. The rate of development of pig embryos. The age of the embryos in hours is defined as 0 at the moment of ovulation. Values are given as the mean + SE of the natural logarithm of the number of cells. The number of embryos is given in parenthesis. The number of pigs is one except for one mean value in which the number of pigs is two as indicated in parenthesis below the mean value.

DISCUSSION

In pigs, it has been reported that the LH peak and the onset of oestrus coincide (Niswender et al., 1970). Other authors, however, reported LH peaks from 2h before to 22h after the onset of oestrus (Tilton et al., 1982) or 8 and 32h before the onset of oestrus in 50% of the animals (Ziecik et al., 1982). These different findings may be explained by the different pig races, the different number of animals or the different methods of blood-sampling and oestrus detection procedures.
In this study LH peaks have been found between 20h before to 32.5h after the onset of oestrus using one type of pigs (Dutch Landrace) with a bloodsampling and oestrus detection frequency every 6 hours. The time interval between the LH peak and ovulation proved to be rather constant in our study (42+5h).

In sheep, the LH peak occurs usually some hours after the onset of oe-strus. LH peaks occurred between 4-16h after the onset of oestrus in a study of Goding et al., (1969) and between 2-9h in a study of Pant et al., (1976). Cumming et al., (1973) reported even larger variations namely LH peaks from 8h before to 9.5 h after the onset of oestrus. Lishman et al., (1974) re-ported LH peaks before oestrus in 22 out of 92 cases. The interval between the onset of oestrus and the LH peak seems to be dependent on race (Quirke et al., 1979), age and previous hormone treatment (cronolone) (Cumming et al., 1973). In this last study it was also shown that the interval between LH peak and ovulation (22-26h) was not dependent on age or cronolone treatment, thus explaining the large variation between the onset of oestrus and ovulation (0.5-35h).

In cattle, the LH peak also occurs usually some hours after the onset of oestrus. Randel et al., (1973) give a mean value of 6h and further an interval between LH peak and ovulation of 24h.
As in pig and sheep, large variation between the onset of oestrus and the LH peak have been reported in cattle. Christensen et al., (1974) give a mean value of 7h for the LH peak to occur after the onset of oestrus with a variation of 0.5-18h. As in pigs and sheep the time interval between the LH peak and ovulation was rather constant in this study. The mean time interval was 24h with a variation between 21-26h. Bernard et al., (1983) also report a large variation in the time interval between the beginning of oestrus and the beginning of the LH peak (-8 to +9h) and a less variable interval between the beginning of the LH peak and ovulation (25-29h). The study of Staigmiller et al., (1982) is interesting as it reports that animals on low feed level show a larger variation in the time interval between the beginning of oestrus and the LH peak (mean 7.7h, range 1-16h) than animals on a high feed level (mean 3.7h, range 2-6h).

It is evident from the above cited literature and our own results that the timing of the LH peak varies considerably with respect to the onset of oestrus in pigs, sheep and cattle. The time interval between the LH peak and ovulation is rather constant in these species and the interval between the onset of oestrus and ovulation will therefore also vary considerably. Timing the moment of artificial insemination with respect to the onset of oestrus may therefore result in inseminations at an inappropiate time with respect to ovulation. Our results in pigs suggest that there is an optimal period of insemination between 12 to 28h before ovulation. Increasing the interval between insemination and ovulation results in either conceptions with a high incidence of embryonic mortality or no conception at all. These results are in agreement with those of Dziuk (1970), who reported that the optimum time of insemination is 12h before ovulation in pigs and sheep with induced ovulations. The time interval between the onset of oestrus and ovulation in our study varies between 18 and 76h. This may explain the large between animal variation in embryonic development if the age of the embryos is calculated with respect to the onset of oestrus. An embryo collected e.g.

124

96h after the onset of oestrus may in fact have a "true" age between 20 and
78h. Our results suggest that the development of pig embryos between 60 and
130h after ovulation can be assessed by assuming that the pig embryo doubles
its number of cells every 11h.

ACKNOWLEDGEMENTS

The authors are indebted to Dr. G.D. Niswender for the rabbit anti-
porcine LH serum (batch 566) and Dr. L.E. Reichert for the purified LH (LER
786-3). The personnel of the ILOB and the members of the working party on
"early pregnancy" are acknowledged for their contribution of this study.
The LEB funds are acknowledged for subsidizing part of the hormone analyses
in this study.

REFERENCES

Bernard, C., Valet, J.P., Béland, R. and Lambert, R.D. 1983. Prediction of
    bovine ovulation by a rapid radioimmunoassay for plasma LH. J. Reprod.
    Fertil., 68, 425-430.
Christensen, D.S., Hopwood, M.L. and Wiltbank, J.N. 1974. Levels of hor-
    mones in the serum of cycling beef cows. J. Anim. Sci., 38, 577-583.
Cumming, I.A., Buckmaster, J.M., Blockey, M.A. deB., Goding, J.R., Winfield,
    C.G. and Baxter, R.W., 1973. Constancy of interval between luteinizing
    hormone release and ovulation in the ewe. Biol. Reprod., 9, 24-29.
Dziuk, P. 1970. Estimation of optimum time for insemination of gilts and
    ewes by double-mating at certain times relative to ovulation. J. Re-
    prod. Fertil., 22, 277-282.
Goding, J.R., Catt, K.J., Brown, J.M., Kaltenbach, C.C. Cumming, I.A., and
    Mole, B.J. 1969. Radioimmunoassay for ovine luteinizing hormone, secre-
    tion of luteinizing hormone during estrus and following estrogen ad-
    ministration in sheep. Endocrinology, 85, 133-142.
Helmond, F.A., Simons, P.A., Hein, P.R. 1980. The effects of progesterone
    on estrogen-induced luteinizing hormone and follicle-stimulating hor-
    mone in the female rhesus monkey. Endocrinology, 107, 478-485.
Hunter, R.H.F. 1967. Porcine ovulation after injection of human chorionic
    gonadotrophin. Vet. Rec., 81, 21-23.
Liptrap, R.M., Raeside, J.I. 1966. Luteinizing hormone activity in blood
    and urinary oestrogen excretion by the sow at oestrus and ovulation.
    J. Reprod. Fertil., 11, 439-446.
Lishman, A.W., Stielan, W.J., Dreosti, I.E., Botha, W.A., Stewart, A.M. and
    Swart, C.E. 1974. The release of LH at oestrus in ewes on two planes
    of nutrition during lactation. J. Reprod. Fertil., 41, 227-230.
Niswender, G.D., Reichert, L.E, and Zimmerman, D.R. 1970. Radioimmunoassay
    of serum levels of luteinizing hormone throughout the estrous cycle in
    pigs. Endocrinology, 87, 576-580.
Pant, H.C., Hopkinson, C.R.N. and Fitzpatrick, R.J. 1977. Concentration of
    oestradiol, progesterone, luteinizing hormone and follicle-stimulating
    hormone in the jugular venous plasma of ewes during the oestrus cycle.
    J. Endocr., 73, 247-255.
Quirke, J.F., Hanrahan, J.P. and Gosling, J.P. 1979. Plasma progesterone
    levels throughout the oestrus cycle and release of LH at oestrus in

sheep with different ovulation rates. J. Reprod. Fertil., 55, 37-44.

Randel, R.D., Short, R.E., Christensen, D.S. and Bellows, R.A. 1973. Effects of various mating stimuli on the LH surge and ovulation time following synchronization of estrus in the bovine. J. Anim. Sci., 37, 128-130.

Staigmiller, R.B., England, B.E., Webb, R., Short, R.E., and Bellows, R.A. 1982. Estrogen secretion and gonadotropin binding by individual bovine follicles during estrus. J. Anim. Sci., 55, 1473-1482.

Tilton, J.E., Foxcroft, G.R., Ziecik, A.J., Coombs, S.L. and Williams, G.L. 1982. Time of the preovulatory LH surge in the gilt and sow relative to the onset of behavioral estrus. Theriogenology, 18, 227-236.

Ziecik, A., Krzymowska, H. and Tilton, J.E. 1982. Porcine LH levels during the estrus cycle, gestation, parturition and early lactation. J. Anim. Sci., 54, 1221-1225.

DISCUSSION

Bouters:

Embryonic oestrogens are high in the sow and mare and are low in other domestic animals. Could this be related to a luteotrophic effect of oestrogens in the sow and a luteolytic effect in ruminants?

Heap:

The idea is attractive but evidence for a luteotrophic effect of oestrogen is lacking in the mare. The sow is the only domestic animal that is known to produce substantial amounts of blastocyst oestrogens and in which oestrogens are known to be luteotrophic.

Bouters:

Based on the electronmicroscopic plates you presented could the cow placenta be classified as epitheliochorialis and the sheep placenta as syndesmochorialis?

Heap:

I do not think that simple classification schemes can be used any longer in the cow and sheep. In both species it is only the maternal uterine epithelium that is modified, and the basement membrane is never penetrated as in the classical view of syndesmochorial placentation.

Sreenan:

Is there evidence of trophectoderm cell invasion in the cow as you described for the sheep and if so when does it occur?

Heap:

For the sheep the evidence suggests that it occurs about Day 13; for the pig perhaps Day 11 or 12. It has not yet been described for the cow.

Fischer:

Can you speculate on the significance of the maternal signal you mentioned, viz., the oxytocin suppression?

Heap:

The absence of an increase in uterine oxytocin receptors at Day 16 of pregnancy presumably means that oxcytocin induced PGF-2α secretion by the endometrium is impaired. While ovarian oxytocin secretion continues during the first two weeks of pregnancy (Flint and Sheldrick, 1983) it fails to activate uterine PGF-2α pulsatile secretion of the high amplitude type.

Courot:

What is the duration of sensitivity to progesterone monoclonal antibodies in the early mouse embryo?

Heap:

The evidence is that 2-4 cell embryos are most sensitive to progesterone depletion.

Wilmut:

Is embryo transport in the oviducts affected by progesterone antibody treatment?

Heap:

In most instances, tubal transport is unaffected.

Helmond:

You showed that the number of oxytocin receptors increased in cycling sheep and remained basal in pregnant sheep. Progesterone level seems to decline earlier than the increase in oxytocin receptors. How does this fit your model of C.L. regression?

Heap:

Further information is required to establish the temporal relationship between the generation of the oxytocin receptor

population and the first decline in circulating progesterone
concentration.

Robinson:

Is the arrested development of the mouse embryos due to a
lower concentration of free progesterone or to it's total
absence?

Heap:

We are currently developing a technique to measure free
circulating progesterone level after antibody treatment.

Boland:

Can the embryos from mice which have been exposed to the
progesterone antibody treatment be rescued?

Heap:

Administration of exogenous progesterone simultaneously
with the antibody treatment will reverse the antifertility
effect.   A similar result is obtained with progesterone
administered up to 48 hours after antibody treatment though
embryo survival is less.   After 48 hours progesterone is
ineffective in reversing the antibody effect.   Embryo loss
is greatest after antibody treatment at 32 hours after mating
if excess exogenous progesterone is not given simultaneously.

Robinson:

Would you expand on the role of prolactin in EPF activity?

Koch:

After replacement of the pituitary in the ovary-oviduct
culture system, with prolactin, FSH or LH, only the prolactin
stimulated EPF production.   In in vivo experiments EPF
disappears within 48 hours of hypophysectomy but it's
production can be maintained by daily administration of 2 X
10 μg prolactin.

Wilmut:

Does prolactin crossreact with EPF?

Koch:

No. See Wilson et al. (1984) J. Reprod. Immunol. 6, 253.

Helmond:

Prolactin is important for maintenance of C.L. function and for EPF production in rodents. In farm animals it is not essential for C.L. maintenance; does this exclude a role for prolactin in EPF production in farm animals?

Koch:

We do not know.

Sreenan:

The data of Ayalon (1978) suggests that uterine electrolytes as well as proteins may affect embryo survival. Have you any data or comment on this?

Fischer:

To date we have no measurements on electrolytes but I agree that the results of Ayalon (1978) should be investigated further.

Wilmut:

Can uterine environment modify the rate of embryonic development in the absence of exogenous hormones?

Fischer:

Uterine flushing composition differs between individuals in the absence of hormonal treatment but I am not sure of the effect of composition on embryo development.

Courot:

Could the tolerance for embryo-uterine asynchrony be related to the duration of the "free-life" of the embryo in utero or to the duration of normal pregnancy?

Fischer:

The preimplantation periods for sheep and cattle are about 18 and 35 days respectively.   Embryo transfer experiments have shown that asynchrony in these species should not exceed about ± 2 days, this would not suggest any direct relation-ship between preimplantation period and tolerance for asynchrony.

Greve:

Which is the more important donor-recipient or embryo-recipient synchrony?

Fischer:

Certainly for the rabbit embryo-recipient synchrony is the more important.

Sreenan:

Definition of embryo-maternal synchrony is more difficult for the cow.   For example on Day 7, embryos ranging in stage from early morula to expanded blastocyst may be recovered from any one donor.   In this situation synchrony is best defined in terms of embryo stage and recipient cycle stage.

Bouters:

It has been postulated that in cattle embryo transfer a closer degree of synchrony is required for Day 7 - 8 than for Day 13-14 embryos.   Is there evidence for this?

Sreenan:

There are insufficient data available for Day 13-14 embryo transfers to draw such a conclusion.

Heap:

Have you identified whether oestradiol-17β is metabolised by morulae and blastocysts in culture and do you think the

early embryo is exposed to a high concentration of oestradiol-
17B   in vivo?

Niemann:

    Our in vitro studies suggest that is is not metabolised
by the morulae or blastocysts.   Other groups have shown the
presence of a high concentration of oestrogens in uterine
fluid at this stage.

Greve:

    Did you inject pigs with either progesterone or oestrogen
and measure the rate of embryonic development?

Niemann:

    Increasing amounts of progesterone administered to
gilts for the first four days after mating had little effect
on embryo development.   Embryo development was only
inhibited with excessively large doses of progesterone.
Excess progesterone seemed to be metabolised in the liver
and endometrium.   We do not have data on the effect of
oestradiol.

Bouters:

    Is it likely that the oestradiol-17B is metabolised to
oestrone?

Niemann:

    We have to no evidence to suggest this.   Perhaps
oestrogens are necessary to switch on some genes regulating
further embryonic development.

Heap:

    How do you measure embryo development?

Niemann:

    Development was defined as the change in stage from
morula to blastocyst or hatched blastocyst.   Frequently cell
numbers were counted.

Fischer:

Do you have any information on the effect of gamete age on the subsequent sex-ratio?

Helmond:

No.

Boland:

How was the timing of ovulation measured relative to the LH peak and or to the onset of oestrus?

Helmond:

Time of ovulation was defined relative to the time at which progesterone reached 1 ng per ml above basal level. This resulted in an estimated interval of 42 hours between LH peak and ovulation.

Bolet:

Did delayed timing of fertilisation relative to the time of ovulation increase the incidence of embryonic mortality due to chromosomal abnormalities?

Helmond:

Too few embryos were collected after delayed fertilisation to be conclusive. Indeed the overall incidence of chromosomal abnormalities in pig embryos was very low.

Fischer:

Do environmental influences such as frequent blood sampling affect the interval between the onset of oestrus and LH peak?

Helmond:

In our situation this was not the case.

Courot:

Did you measure ovulation rate and was there any

relationship between ovulation rate and the interval between onset of oestrus and LH peak?

Helmond:

We did not find such a relationship but it seems to exist for sheep.

Greve:

Did you find a relationship between oestrogen level and oestrous intensity or duration?

Helmond:

No.

Greve:

What was the interval between oestrogen and LH peaks?

Helmond:

The oestradiol peak precedes the LH peak by a fairly constant six hours.

Greve:

Could urinary oestrogen be used as an indicator of the time of ovulation by predicting the timing of the LH surge?

Helmond:

Because of the constant interval between the oestrogen and LH peaks this would seem possible.

Koch:

How did you define onset of oestrus?

Helmond:

Oestrous onset was defined as the mid-point between the time of the last oestrous check and the first mating by a vasectomised boar. Oestrous checks were carried out every six hours.

Courot:

What is the optimum time for A.I. in the pig under field conditions?

Helmond:

From our studies we found that ovulation occured at the end of oestrus.   We also found that both the duration of oestrus and the time interval between oestrous onset and LH peak were highly repeatable within animals.   By measuring the duration of oestrus in one cycle this allows insemination to be carried out at the optimum time of 12-24 hours before the end of oestrus in a following cycle.

Because, however, we found that a long duration of oestrus was associated with a delayed LH peak and ovulation, we advise that such sows be inseminated two or three times.

Robinson:

In our laboratory we observed a biphasic LH peak in 9 of 36 sheep in a recent experiment and this was associated with a higher than average level of embryo mortality. Is there any evidence for this in your studies?

Helmond:

We did not find any biphasic LH peaks in the pig.   In the case you refer to perhaps the first LH surge initiated premature maturation of the oocytes which resulted in too long a maturation period in the follicle before ovulation was induced by the second LH surge.

THE INFLUENCE OF PROGESTERONE PROFILE

ON EMBRYO SURVIVAL IN EWES

I. Wilmut, C.J. Ashworth and D.I. Sales*

A.F.R.C. Animal Breeding Research Organisation,
Dryden Laboratory, Roslin, Midlothian U.K.
*A.F.R.C. Poultry Research Centre, Roslin,
Midlothian U.K.

ABSTRACT

Changes in the progesterone profile after mating are critical for the establishment of pregnancy in sheep. Studies in ovariectomised ewes have shown that low pre-luteal levels are required from immediately after oestrus until the level increases to luteal phase levels on Day 4 or 5. In intact, mated ewes, differences in embryo survival were associated with variation in the progesterone profile. Embryo survival was lower in ewes with lower levels of progesterone immediately after oestrus. Differences in progesterone level accounted for the seasonal effect on survival, and for part of the effect of number of ovulations. These observations create new opportunities for treatment with progesterone to increase embryo survival in sheep and cattle.

INTRODUCTION

It has long been known that progesterone secreted by the corpus luteum is essential for the establishment of pregnancy. In recent years, two approaches have been used to define the role of such progesterone. In the first, progesterone was supplied to ovariectomised ewes in order to establish the profile required for the establishment of pregnancy. In the second, the progesterone profile was measured in mated ewes and a comparison made between the profile when all of the embryos survived and that when some or all of the embryos did not survive. The observations made in these experiments will be reviewed, and their implication for methods of improving embryo survival considered.

THE REQUIREMENT FOR PROGESTERONE

Three conclusions have been drawn from experiments in which progesterone has been given to ewes. First, the time of changes in the uterus is controlled by the time of the increase in progesterone concentration to levels typical of the luteal phase. Second, pregnancy will only occur if a certain minimal level of progesterone is present during the luteal phase. Third, a lower, but equally important, level of peripheral progesterone is required from oestrus until the time of the increase to luteal levels.

The failure to establish pregnancy if embryos are transferred between animals that were not on heat at the same time shows that uterine secretions change such that embryos must develop in a uterus that is at an appropriate stage. In sheep, if embryos are out of phase during the second week of pregnancy, their development becomes abnormal (Wilmut and Sales, 1981) and as they fail to inhibit luteolysis they are expelled from the uterus as the ewe returns to oestrus (Lawson, Parr and Cahill, 1983). Two experiments have shown that the timing of the increase in progesterone concentration controls the timing of changes in uterine function. When progesterone injections (25 mg/day) were given for four days from the day of oestrus the oestrous cycles were shortened to 13.8 ± 0.8 days compared with 17.5 ± 0.3 days in control ewes (Lawson and Cahill, 1983). If Day 10 embryos were transferred to such treated ewes on Day 10 of the cycle, no pregnancies occurred. By contrast, 8 of 12 recipients became pregnant when such embryos were transferred to treated ewes six days after oestrus. Apparently the treatment with progesterone had advanced the time of changes in the uterine environment. A similar effect was observed in ovariectomised ewes given injections of increasing amounts of progesterone designed to mimic the changes in intact animals (Miller and Moore, 1976). After treatment according to the standard regimen, the greatest proportion of pregnancies was obtained if the Day 4 embryos were transferred to Day 4 recipients. By contrast, when essentially the same regimen was given 1.5 days earlier than in the control ewes, maximum survival was obtained if Day 4 embryos were transferred to recipients 2.5 days after oestrus. In both experiments, the advance in the time of changes in the uterus was apparently the same as the interval by which the progesterone administration had been advanced.

The proportion of pregnancies that become established depends upon the level of progesterone present during the luteal phase. Different doses of progesterone have been given either to ewes that were ovariectomised after mating or which received embryos by transfer. The number that had normal embryos on Day 20 increased from 0 of 14 given 1 mg/day to 2 of 12 ewes given 4 mg/day and 11 of 14 ewes given 16 mg/day (Bindon, 1971). In similar ewes given 5, 10, 15, 20 or 25 mg/day, embryo survival to Day 21 was 69%, 83%, 79%, 90% and 82% respectively (Parr, Cumming and Clarke, 1982). In other experiments, a majority of ewes given 10-20 mg/day of progesterone to mimic the luteal phase were found to become pregnant (Moore and Rowson, 1959; Trounson and Moore, 1974). These observations suggest that very few ewes can remain pregnant if they are given 4 mg/day of progesterone or less, whereas pregnancy will be established in a majority of ewes given 10 mg/day or more. As the effect of doses between 4 and 10 mg/day is not known, the critical threshold has not been defined.

A lower level of progesterone must be present until the time of the rise to levels typical of the luteal phase or very few pregnancies will occur. This effect became apparent when embryos were transferred to ovariectomised ewes on Day 6, after sub-cutaneous insertion of silastic implants containing progesterone on Day 4 or 5. Among ewes receiving injections of progesterone until the time of implant insertion 9 of 13 become pregnant; by contrast, only 3 of 16 conceived if they received no injections of progesterone. There is no information on the optimum level of progesterone during the period immediately after oestrus, but it seems that the amount produced by the adrenal gland is not adequate for the subsequent establishment of pregnancy.

Together these observations show that there are at least two phases in the control of uterine function by the progesterone secreted after mating. A period from oestrus until Day 4 or 5 when the level should be low, followed by an increase to levels typical of the luteal phase. While this general pattern has been observed in many ewes, no study had been carried out to define the variation between ewes or between different cycles in the same animal. An experiment has been carried out to define this variation and to compare the profiles in ewes in which all embryos survived with those in which some embryos were lost (Ashworth, 1984; Ashworth, Sales and Wilmut, 1984).

PROGESTERONE PROFILES AND EMBRYO SURVIVAL

Twenty eight ABRO Damline ewes were used in this experiment. Ewes were housed indoors during periods of blood collection lasting for approximately 3 weeks around the time of mating. Oestrous cycles were synchronised by two intramuscular injections each of 100ug of cloprostenol (Estrumate, I.C.I. plc, Macclesfield, Cheshire) given 9 days apart. At the second oestrus following synchronisation, the ewes were hand mated by at least two rams from a group of seven. Blood samples were obtained at 7 am, 2 pm, and 9 pm for 7 days around the time of mating, and subsequently twice daily, at 8 am and 4 pm until Day 16, when the ewes were turned out

into a paddock.   The number of ovulations was determined during a
laparotomy either on Day 30 after mating or, if the ewes returned to
oestrus, within two days of that oestrus.   The number of foetuses was
recorded in animals that did not return to oestrus.   During the surgery
the reproductive tract was handled with great care and 10 ml sterile
saline was introduced into the peritoneal cavity in order to minimise
the adhesions.   Following surgery, 100ug cloprostenol was given to
pregnant ewes in order to induce abortion.

This procedure was repeated on some of the ewes during September,
November, January and March, but no ewe was studied on more than three
occasions.   Ewes were allowed at least one cycle between successive
sampling periods.   Blood samples were collected on three occasions from
16 ewes and on two occasions from 12 ewes.   Details of the assay for
luteinising hormone (LH) and progesterone have been presented elsewhere
(Ashworth, 1984).   The plasma samples were assayed from those ewes
which experienced embryonic loss and from those ewes that were studied on
three occasions; a total of 48 cycles from 18 ewes.   Progesterone
estimates were log-transformed and a smoothed progesterone profile
calculated; given three consecutive log progesterone values X1, X2 and X3,
the smoothed value of X2 was (0.25 X1 + 0.50 X2 + 0.25 X3).   These values
were used in all subsequent calculations.

The profiles were divided into components suitable for statistical
analysis.   The mean baseline value was defined as the mean progesterone
concentration from samples collected on Days 0 and 1.   The mean luteal
level was defined as the mean progesterone concentration in samples
collected on Days 9, 10, 11 and 12.   The time of the increase in
progesterone concentration was assessed by determining the time taken for
progesterone concentrations to reach 25%, 50% and 75% of their mean luteal
level (T25, T50, T75).   The time when the first of 2 consecutive
progesterone values were greater than 0.5 ng/ml was defined as T0.5ng.
These times were all described in relation to the peak in LH.   Binomial
survival data were analysed using a linear model with a logit link
function.

There was considerable variation between ewes in both circulating
progesterone concentrations and in the concentration associated with the
survival of all of the embryos.   As there were significant differences
between animals in the proportion of embryos surviving (p<0.01), separate
within- and between-animal analyses were performed.   In the within-animal

analysis, partial or total embryo loss was associated with lower baseline progesterone levels (p<0.05). There was also a tendency for lower luteal levels on these occasions (p<0.10). There was no significant association between any of the four timing components of the progesterone profile and the probability of an embryo surviving.

The proportion of individual embryos surviving decreased as the number of ovulations increased (p<0.01). As the number of ovulations increased the level of progesterone during the luteal phase also increased, but each additional corpus luteum produced less progesterone (p<0.01). In the statistical anaylsis, differences in luteal phase progesterone accounted for part of the effect of number of ovulations on embryo survival. This effect could indicate that each embryo requires a certain amount of progesterone and that the additional production by the extra corpora lutea was insufficient. The number of ovulations did not significantly affect the other progesterone parameters measured.

Embryo survival was affected by the season of mating. Survival was lower in March than in any other month (p<0.05). When the effects of progesterone were included in the statistical model the effect of season was no longer significant, indicating that the effect of season is mediated entirely by progesterone. Analysis of variance revealed lower mean baseline progesterone levels following mating in March (p<0.01). There was no effect of season on the five other progesterone parameters.

Repeatabilities of mean baseline and mean luteal progesterone values were 0.154 ± 0.002 (S.E.M.) and 0.171 ± 0.002 (S.E.M.) respectively.

These observations are the first to show an association between progesterone profile and embryo survival in healthy, well managed ewes. This effect accounted for seasonal variation in embryo survival, and for some of the effect of variation in the number of ovulations. As well as providing an explanation of the mechanisms by which some environmental factors influence embryo survival, this effect may also account for some of the previously unexplained loss.

While the association observed in this experiment may reflect a direct influence of progesterone on embryo survival, there are alternative explanations. It may be that follicles that develop into corpora lutea that produce smaller amounts of progesterone than normal are also unable to support oocyte maturation. It is possible that the early embryo has luteotrophic effects and that they stimulate the ewe to produce more progesterone. As the different levels were observed only hours after ovulation this seems unlikely. Differentiation between these suggestions may be possible by administration of progesterone. If exogenous progesterone increases the proportion of embryos that survive, the embryos could not be abnormal and it would seem most likely that the differences observed in this experiment were due to inadequate levels of progesterone.

An association between the proportion of pregnancies and progesterone level on Day 1 to Day 3 after oestrus has also been observed in cattle (Lee and Ax, 1984), and it now seems probable that similar mechanisms may control embryo survival in the two species.

IMPLICATIONS

These observations create new opportunities for the development of methods of pharmacological intervention to increase embryo survival in sheep and cattle. Similarly, they reveal potential methods of "synchronising" recipients for embryo transfer with the additional advantage of a potential improvement in the proportion of pregnancies.

Treatment with exogenous progesterone during the luteal phase has resulted in increased embryo survival in cattle (Sreenan and Diskin, 1983). In view of the present results, application of smaller amounts of progesterone from Day 1 to Day 4 inclusive, before treatment during the luteal phase, may have a greater effect. Care would be needed to ensure that the amount of progesterone given immediately after oestrus was not sufficient to mimic that in the luteal phase. In sheep, 10 mg/day would be too much, but the optimum dose is not known.

Exogenous progesterone may also be used to control the time of changes in uterine function in recipient cattle as discussed elsewhere (Wilmut, Sales and Ashworth, 1985). In this case it would be necessary to give amounts of progesterone equivalent to that found during the luteal phase during the first four days after oestrus. Transfer would have to be carried out in relation to the time of this treatment. Such treatment would control exactly the appropriate time for embryo transfer and could also lead to an increase in the proportion of pregnancies.

The estimates of repeatability of around 0.15 for the two components of the progesterone profile, suggest that genetic selection for this character would have only a small effect. It seems that the profile is influenced greatly by environmental factors, despite the fact that it is evidently important in the establishment of pregnancy.

The present review has considered only the progesterone profile
after mating.   Progesterone and oestradiol before mating are both known
to be essential for the establishment of pregnancy (Miller and Moore,
1976).   There is little information available on the variation in levels
of these hormones in intact animals or on the influence of this
variation on the subsequent pregnancy.   While new opportunities for
treatment are available, much remains to be learned.

REFERENCES

Ashworth, C.J. 1984. Maternal Factors Affecting Early Pregnancy in Sheep.
    Ph.D. Thesis, University of Edinburgh.
Ashworth, C.J., Sales, D.I. and Wilmut, I. 1984. Patterns of progesterone
    secretion and embryonic survival during repeated pregnancies in
    Damline ewes. Proc. 10th Int. Congr. Anim. Reprod. and A.I.,
    Illinois 2, 74.
Bindon, B.M. 1971. The role of progesterone in implantation in the
    sheep. Aust. J. Biol. Sci., 24, 149-158.
Lawson, R.A.S. and Cahill, L.P. 1983. Modification of the embryo maternal
    relationship in ewes by progesterone treatment early in the
    oestrous cycle. J. Reprod. Fert., 67, 473-475.
Lawson, R.A.S., Parr, R.A. and Cahill, L. 1983. Evidence for maternal
    control of blastocyst growth after asynchronous transfer of
    embryos to the uterus of the ewe. J. Reprod. Fert., 67, 477-483.
Lee, C.N. and Ax, R.L. 1984. Milk progesterone of dairy cows injected
    with gonadotrophin releasing hormone at the first post partum
    breeding. Proc. 10th Int. Congr. Anim. Reprod. and A.I., Illinois
    2, 401.
Miller, B.G. and Moore, N.W. 1976. Effects of progesterone and oestradiol
    on RNA and protein metabolism in the genital tract and on survival
    of embryos in the ovariectomised ewe. Aust. J. Biol. Sci., 29,
    565-573.
Moore, N.W. and Rowson, L.E.A. 1959. Maintenance of pregnancy in
    ovariectomised ewes by means of progesterone. Nature (London)
    184, 1410-1411.
Parr, R.A., Cumming, I.A. and Clarke, I.J. 1982. Effects on maternal
    nutrition and plasma progesterone concentrations on survival
    and growth of the sheep embryo in early gestation. J. agric. Sci.,
    Camb. 98, 39-46.
Sreenan, J.M. and Diskin, M.G. 1983. Early embryonic mortality in the
    cow: its relationship with progesterone concentration. Vet. Rec.
    112, 517-521.
Trounson, A.O. and Moore, N.W. 1974. Effect of progesterone and
    oestrogen on the survival and development of fertilised ova in
    the ovariectomised ewe. Aust. J. Biol. Sci., 27, 511-517.
Wilmut, I. and Sales, D.I. 1981. Effect of an asynchronous environment
    on embryonic development in sheep. J. Reprod. Fert., 61, 179-184.
Wilmut, I., Sales, D.I. and Ashworth, C.J. 1985. The influence of
    variation in embryo stage and maternal hormone profiles on embryo
    survival in farm animals. Theriogenology. (In press).

PROGESTERONE AND EMBRYO SURVIVAL IN THE COW

M.G. Diskin, J.M. Sreenan

The Agricultural Institute,
Belclare, Tuam, Galway,
Ireland

ABSTRACT

Embryonic mortality accounts for the major portion of reproductive failure in all domestic animals. Progesterone, secreted by the Corpus Luteum is essential for the establishment and maintenance of pregnancy. Progesterone level in the cycle preceeding insemination does not seem to affect embryonic survival. Following insemination, progesterone level increases in pregnant cows with most evidence suggesting a divergence as early as Day 10-12 between pregnant and non-pregnant cows. While it is presumed that this increase arises from a luteotrophic embryonic signal there is not direct evidence to substantiate this before Day 16. Progesterone supplementation has increased embryo survival in both normal and repeat breeder cows though the number of animals are limited in individual experiments. The administration of hCG during the luteal phase significantly increased the number of corpora lutea and the circulating progesterone level but not embryo survival rates.

INTRODUCTION

Embryonic mortality accounts for the major portion of reproductive failure in all domestic species. In cattle up to 40% of fertilised ova may die. Studies based on ovum, embryo and foetal recoveries at various stages after breeding, indicate that early embryonic death between Days 8 and 18 account for 75-80 percent of all embryonic and foetal mortality (Diskin & Sreenan, 1980; Roche et al.1981). Because most losses occur before Day 18 the normal sequence of luteolytic events is not affected and oestrus and ovulation recur within an interval of 18-25 days. While this gives a further opportunity for rebreeding and pregnancy establishment it nevertheless causes a substantial loss in production.

The extent of embryonic loss is established but it's cause are poorly understood. It has been suggested (Bishop, 1964) that such losses are unavoidable and represent nature's way of eliminating unfit genotypes at an early stage.

Recent evidence in cattle (Gayerie, 1983) would suggest,
however that chromosonal abnormalities are not major causes
of embryonic loss.   Embryo transfer experiments (Rowson et.
al., 1969;   Sreenan, 1977;   Kunkel & Stricklin, 1978) have
highlighted the importance of close synchrony between donor
and recipient to achieve high pregnancy rates.   The recent
sheep studies of Wilmut and Sales (1981) Lawson & Cahill (1983)
and Lawson et al. (1983) have focussed on the importance of
embryo-uterine asynchrony as a major cause of embryonic
mortality.   Wilmut and Sales (1981) and Lawson et al.(1983)
showed that while the rate of embryonic development could be
accelerated by the transfer of 'younger' embryos to a
physiologically 'older' uterus, this would eventually prove
fatal for the embryo.   Lawson et al (1983) concluded that
such 'abnormal' embryos were also unable to prevent luteolysis.
The physiological state of the uterus can be modified by
progesterone supplementation early in the cycle (Lawson &
Cahill, 1983).   These transfer experiments suggest that
asynchrony between the embryo and uterine environment is a
likely cause of embryonic death and that some of this
asynchrony arises spontaneously.

In domestic animals, progesterone secreted by the corpus
luteum is essential for the establishment and maintenance of
pregnancy.   Because of it's indespensible role, many
researchers have attempted to relate progesterone concentrations
in plasma, serum or milk, during the oestrous cycle preceeding
breeding and during the early and mid-luteal phases after
breeding to conception rate.   Others have used either direct
progesterone supplementation or luteal stimulation to
increase progesterone levels in order to increase pregnancy
rates.

THE EFFECT OF PROGESTERONE LEVEL PRECEEDING SERVICE ON
SUBSEQUENT CONCEPTION RATE

The documented effects of progesterone level during the
oestrous cycle preceeding insemination on conception rate
are inconsistent and often contradictory.

An inverse relationship between progesterone level during

the two day period before oestrus and the number of blastomeres in the ovum three days after breeding has been reported (Henricks, et al. 1971). These authors concluded that high levels of progesterone before oestrus could adversely affect ovum development and, therefore, subsequent embryo survival rates. However, these observations are in disagreement with those of Corah et al. (1974) who observed higher concentrations of progesterone 3 days before breeding in cows that conceived. Similarly, Folman et al. (1973) report a positive correlation between progesterone levels at 2, 7, 8, 11, 12, and 15 days before insemination and conception rate. Holness et al. (1977) observed significantly higher milk progesterone levels 7-9 days before insemination in cows that conceived and also a positive correlation (r=.73) between the mean of two samples before insemination and the mean of six samples in the subsequent luteal phase.

In an Israeli study, investigating the cause of reduced fertility in dairy herds in summer, Rosenberg et al. (1977) found that progesterone levels in cycles preceeding fertile inseminations reached a peak near the time of insemination, while in cows that did not conceive began to decrease as early as 8-11 days before insemination. They also recorded a positive correlation (r= 0.49) between the difference in plasma progesterone concentration at 4-7 days and 8-11 days before insemination and subsequent conception rate. However, such differences or correlations were not evident during the luteal phase preceeding insemination in cows bred during the winter.

In a later study, Bulman and Lamming (1978) found no relationship between pre-service progesterone levels and conception rate.

In a study at this laboratory differences in the progesterone profiles of cows conceiving and cows failing to conceive were not observed during the oestrus cycle preceeding insemination (Fig. 1).

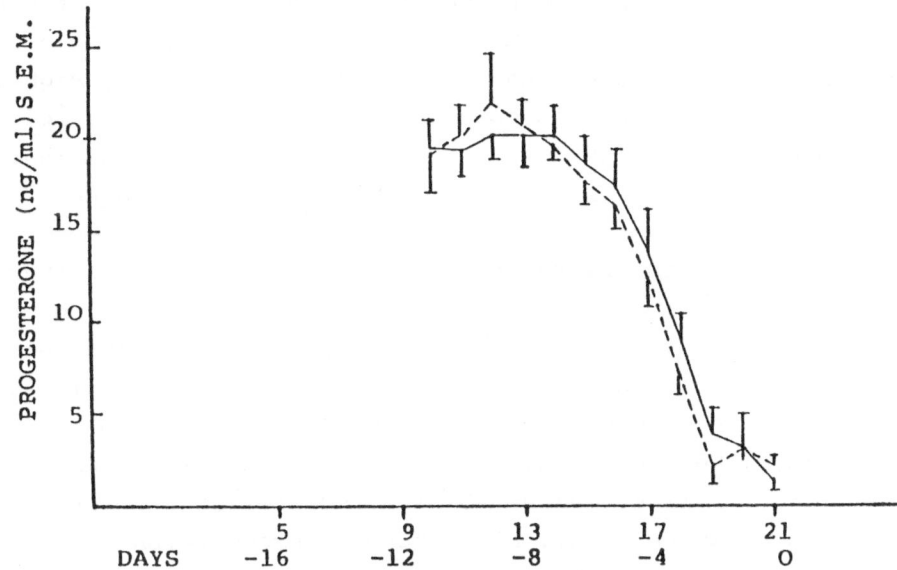

FIG. 1.   Milk progesterone levels during the oestrous cycle
          preceeding insemination in cows that became
          pregnant (---;n=25) and remained open (——;n=25)

It is both difficult to rationalise the differences
between all of these reports and also to decide whether a
deficiency of progesterone during the luteal phase preceeding
insemination or the timing and nature of the progesterone
decline at luteolysis are critical factors in determining
conception rate.

THE RELATIONSHIP BETWEEN PROGESTERONE LEVEL FOLLOWING
INSEMINATION AND CONCEPTION RATE

The increase, levelling off and fall in progesterone
level during the oestrous cycle reflects the development,
maintenance and regression of the corpus luteum. If, after
breeding, fertilisation and normal embryo development occurs,
progesterone remains at about the mid-luteal level indicating
retention of the corpus luteum in a functional state. This
change in corpus luteum lifespan and function, caused by the
presence of the embryo transfer (Betteridge et al. 1979;
Betteridge et al. 1984) and by the embryo removal and infusion
experiments of Northey and French (1980).

A divergence in plasma progesterone concentration between pregnant and non-pregnant cows during the early and mid-luteal phases has been reported by a number of workers. Higher progesterone levels in cows becoming pregnant might reflect a luteotrophic signal from the embryo (Bulman & Lamming, 1978) alternatively low levels might result in an inadequate uterine environment.

The literature relating to early and mid-luteal phase progesterone levels in pregnant, non-pregnant and non-bred cyclic cows is also contradictory and often inconsistent. A number of authors have failed to establish any relationship between progesterone levels and conception rate up until about Day 16, (Shemesh, et al. 1968; Pope et al. 1969; Robertson & Sarda, 1971; Folman et al. 1973; Hasler et al. 1980).

A positive relationship between progesterone levels during the early luteal phase and pregnancy rate has been observed on Day 3 (Maurer & Echternkamp, 1982) Day 6 (Erb et al. 1976) and Day 7 (Randel et al. 1971). Thompson et al. (1980) observed a faster rise in progesterone level between Days 5-10 in pregnant than in either non-pregnant or 'sham' inseminated heifers. They concluded that the increased progesterone level in pregnant animals was due to a stimulatory effect of the embryo on the corpus luteum. However, the number of animals involved in the experiment was too small to allow firm conclusions to be drawn. There is no indication from in-vivo or in-vitro experiments that the bovine embryo is luteotrophic before Day 10. Erb et al. (1976) and Maurer and Echternkamp (1982) showed that the lower progesterone levels in non-pregnant animals were due to delayed oestrus following progesterone decline and or to an increased interval from the onset of oestrus to the pre-ovulatory LH surge.

However, Ayalon (1973) and Linares et al. (1982) in similar experiments to that of Maurer and Echternkamp (1982) failed to observe a difference in progesterone concentration between Days 0 and 7 in 'normal' and in 'repeat' breeder cows from which normal and degenerating embryos were recovered.

During the mid-luteal phase a higher progesterone level

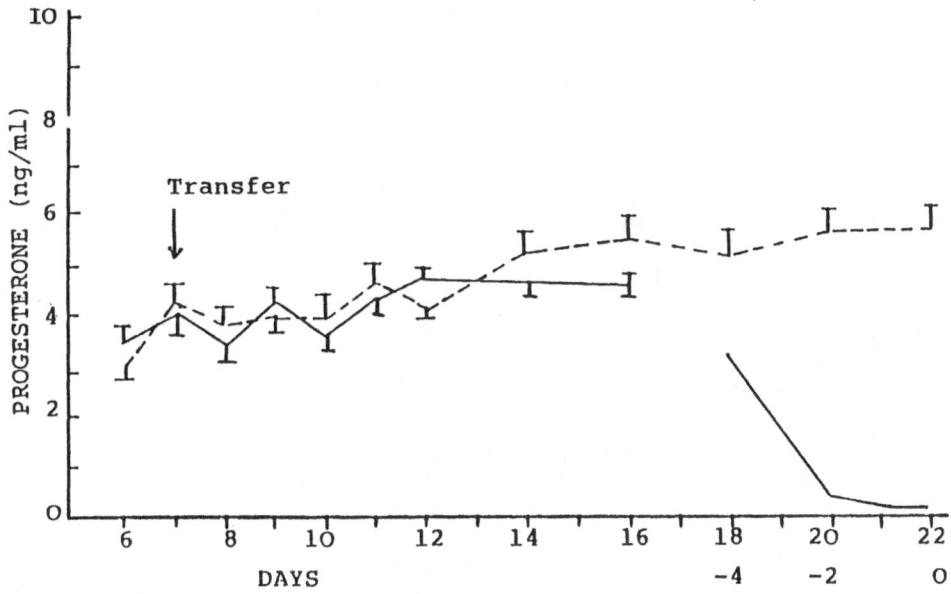

FIG 2.   Plasma progesterone levels in pregnant (———;n=15)
          and non-pregnant (---;n=5) recipients following
          embryo transfer.

in pregnant animals has been observed form Day 9 (Henricks
et al. (1971), Day 10 (Lukaszewska & Hansel 1980;   Henricks
et al. (1970), Days 10-11 and Day 13 (Holness et al. 1977;
Bulman & Lamming, 1978).

        Data from this laboratory (Sreenan & Diskin, 1983) on
plasma progesterone levels in heifers which become pregnant
and those that failed following a single ipsilateral flank
transfer on Day 7 are shown in Fig. 2.   There was no evidence
of a luteotrophic effect immediately following transfer and
differences in progesterone levels between pregnant and non-
pregnant groups did not emerge until Day 16, when the
difference is caused by luteolysis in heifers failing to
become pregnant.

        In a larger study in dairy cows, (Fig. 3) there was no
evidence of a differential rise in progesterone level during
the early luteal phase up to Day 13 in pregnant animals.

148

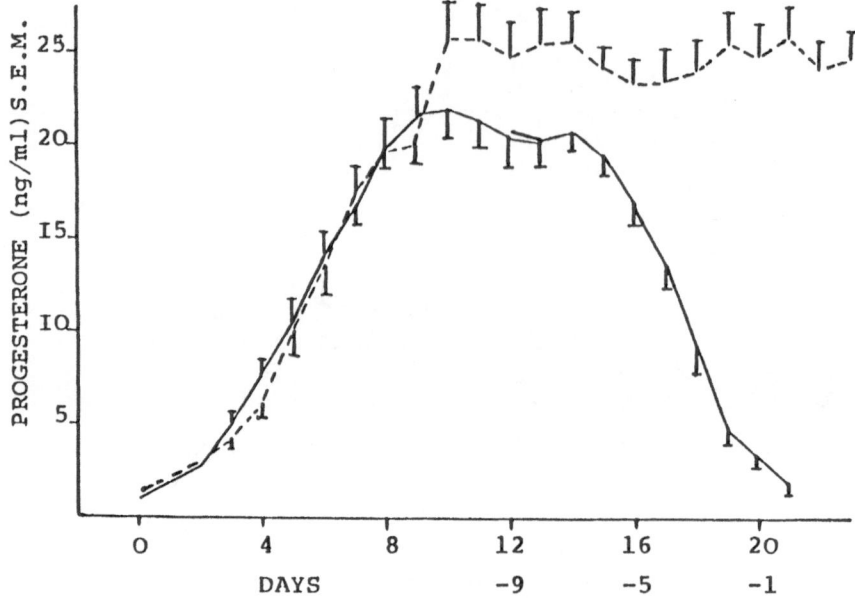

FIG. 3.   Milk progesterone levels after insemination in
pregnant (---;n=50) and non-pregnant cows (——;n=50)

From all of these in-vivo studies, while the trend is
towards an increased progesterone level in pregnant cows, it
is difficult to suggest conclusively that the bovine embryo
is luteotrophic from any specific day onwards.

In-vivo studies have shown that the bovine blastocyst
is luteotrophic at Day 16 (Sreenan et al. 1980).   A
luteotrophic substance has been isolated from cow blastocysts
at Day 15-16 (Shemesh, 1980) and Day 18 (Beal et al. 1981).

Some care must be taken in the assessment, interpretation
and comparison of these in-vivo experiments because of the
frequency of blood sampling relative to the frequency of
naturally occuring fluctuations in progesterone level.   It
must also be borne in mind that the animal numbers in many of
these studies are small and that the experimental,
environmental and management conditions are often quite
different and are thus likely to contribute to variable results.

EFFECT OF PROGESTERONE SUPPLEMENTATION ON CONCEPTION RATE

An insufficient level of progesterone because of luteal dysfunction has been suggested as a cause of early embryonic mortality. In the human, a possible luteal phase defect has been documented by some authors but there remains some confusion in the literature as to it's existence. Certainly the incidence in the human would seem to be very low and the estimates vary between three and eight percent of all infertility cases (Jones and others, 1974; Jones, 1976; Ross, 1979). Progesterone supplementation therapy has been used in the human to allow pregnancy to proceed to the stage where placental steroidogenic function is adequate to maintain pregnancy in the absence of ovarian progesterone.

In the cow, the sole source of progesterone at this time is the corpus luteum. The rationale for progesterone supplementation in the cow is to allow an environment for the embryo to survive to a stage where it's effects (anti-luteolytic and or luteotrophic), are strong enough to maintain a functional corpus luteum. Earlier studies involved progesterone supplementation of sub-fertile or repeat breeding cows. Data from the literature on the effect of progesterone supplementation in repeat breeder cows is summarised in Table 1.

TABLE 1.   Pregnancy rate in 'repeat' breeder cows treated with progesterone

Author	Treatment Schedule	Treatment Groups	
		Control	+Progesterone
Herrick (1953)	500 mg Repositol Prog. Day O	$1/20$ (5%)	$7/20$ (35%)
Dawson (1954)	100 mg Prog. Day 4-5	$3/18$ (17%)	$22/47$ (47%)
Wiltbank et al. (1956)	50 mg Prog. Day 3-34	$12/36$ (33%)	$16/36$ (44%)
Wiltbank et al. (1956)	200 mg Prog. Day 3-34	$8/31$ (26%)	$12/31$ (39%)
Overall	(P 0.05)	$24/105$	$57/134$
% Pregnant		(23%)	(42.5%)

There was a consistent increase in pregnancy rate
following progesterone supplementation in all reports, though
differences were not significant (P> 0.05) in any one study.
When all reports are summarised pregnancy rate is increased
(P< 0.05).   In all of these studies pregnancy rates
recorded in the control cows are very low.   In one study
with cows of normal fertility Johnson et al.(1958) reported
higher (P< 0.05) pregnancy rates in cows that received either
100 mg of repositol progesterone on Days 2, 3, 4, 6 and 9 or
500 mg of progesterone on Day 2.   In that study pregnancy
rates of $26/_{69}$(42%) and $49/_{70}$(70%) were recorded for the
control and combined progesterone supplemented groups
respectively.

The effect of progesterone supplementation on pregnancy
rate is being studied at this laboratory (Table 2).

TABLE 2.   The effect of Progesterone supplementation on
pregnancy rate in heifers

	Treatment Schedule	Treatment Groups	
		Control	+ Progesterone
Experiment 1 pregnancy rate(%)	100 mg Day 10-20	$13/_{25}$ (52%)	$18/_{26}$ (69%)
Experiment 2 pregnancy rate	100 mg Day 5-35	$9/_{20}$ (45%)	$14/_{19}$ (74%)
Total (%)		$22/_{45}$ (49%)	$32/_{45}$ (71%)
% increase in pregnancy rate			+22

Pregnancy rate was increased in both studies by 17 and 29%
respectively though in neither case was the effect significant
(P> 0.05).   However, when both experiments are combined
pregnancy rate is increased (P< 0.05).   Progesterone at
this level decreased luteal tissue weight (P< 0.05).

In a more extensive study the effect of progesterone
supplementation on fertility in dairy cows was examined

over a four year period and the data are shown in Table 3.
Progesterone (100 mg) was admininstered i.m. from Day 10-20.

TABLE 3.   Calving rate (%) in dairy cows supplemented with
            progesterone

Year	Treatment Groups	
	'Control'	+'Progesterone'
1	$19/_{28}$ (67%)	$19/_{24}$ (79%)
2	$22/_{35}$ (63%)	$21/_{31}$ (68%)
3	$24/_{37}$ (65%)	$21/_{33}$ (64%)
4	$24/_{42}$ (57%)	$23/_{42}$ (55%)
TOTAL	$89/_{142}$ (63%)	$84/_{130}$ (65%)

Progesterone supplementation had no effect on fertility
in any one year or in all years combined ($P > 0.05$).   Mean
($\pm$S.D.) repeat intervals for 'control' and 'progesterone
supplemented' cows were 31 ± 13.1 days and 27 ± 12 days
respectively.

Progesterone supplementation did not affect reproductive
performance in the subsequent cycle with calving rates of
$22/_{52}$ (42.3%) and $12/_{36}$ (31.6%) ($P > 0.05$) recorded for 'control'
and + 'Progesterone' cows respectively.   In a further study
with dairy cows progesterone (100 mg) was administered i.m.
earlier (Days 5-10 inclusive) in the luteal phase (Table 4).

TABLE 4.   The effect of progesterone supplementation from
           Day 5-10 on reproductive performance in dairy cows

| | Treatment Groups | | Level of significance |
	'Control'	+'Progesterone	
Calving to service interval ± S.D.	67.3±22.73	69.5±24.4	N.S.
1st service preg. rate(%)	26/65 (40)	29/61 (47.5)	N.S.
Mean repeat interval ± S.D.	22.8±7.6	24.4±12.2	N.S.
2nd service preg. rate(%)	15/35 (42.8)	18/32 (56.3)	N.S.
Services/conception (cow conceiving)	1.63	1.49	N.S.

In this herd, calving rate was low in the 'control' group
but unlike the previously published reports on cows of low
fertility progesterone supplementation did not increase calving
rate (P>0.05).

Progesterone administration had no effect on the
incidence of foetal deaths between implantation and term.
The overall incidence of foetal death was $6/129$(4.6%) and
$7/118$(5.9%) for 'control' and 'progesterone' supplemented
cows respectively (P>0.05).   This gives an overall estimate
of foetal death between implantation and term of 5.4% which is
similar to other published estimates (Boyd et al. 1961).

THE EFFECT OF HCG ADMINISTRATION ON CONCEPTION RATE

As an alternative to progesterone supplementation, hCG
has sometimes been administered as a luteotrophic substance
during the early or mid-luteal phases.   It has also been used
at around the time of luteolysis to supplement the action of
a possible inadequate luteotrophic or anti-luteolytic   signal
from the developing embryo and to, therefore, allow the embryo
to establish as a pregnancy.

Most of the hCG studies report significant increases
(P<0.005) in circulating progesterone level, (Wiltbank et al.

1961; Veenhuizen et al. 1972; Christie et al. 1979; Sreenan
et al. 1979; Greve & Lehn-Jensen 1982; Holness et al. 1982;
Santos-Valadez et al. 1982).   Most of the increase arises
from the induction of accessory corpora lutea, but increased
synthesis from the current corpus luteum has also been
recorded (Santos-Valadez, 1982).

Data from this laboratory relating to the ovarian
response following hCG administered at mid-cycle are shown
in Table 5.

TABLE 5.   The effect of hCG administration (Day 10-20) on
            Corpus Luteum weight and number in pregnant heifers

	Control	+hCG
No. of heifers	22	26
Mean No. CLs	1.04	2.20
Range	1-2	1-4
Mean ($\pm$ S.E.) wt. (g) luteal tissue	5.2 ± 0.4	10.9 ± 1.7

The induction of accessory corpora lutea (2.2) in the
hCG group increased (P<0.001) the total weight of luteal
tissue.

Data on the effect of hCG on pregnancy rate is summarised
in Table 6.

TABLE 6.   The effect of hCG administration on pregnancy rate

| Author | Treatment regime | Pregnancy rate(%) | | (%) |
		Control	+hCG	change
Wiltbank et al.(1961)	1000 i.u.D_{15}-D_{35}	$26/_{41}$(63)	$27/_{39}$(69)	+6
Wagner et al. (1973)	1000 i.u. D_3	$18/_{36}$(50)	$22/_{36}$(61)	+11
	2000 i.u. D_3	$18/_{33}$(55)	$21/_{33}$(64)	+9
Greve & Lehn-Jensen (1982)	1500 i.u.D_{13}-D_{35}	$44/_{59}$(75)	$26/_{32}$(81)	+6
Holness et al. (1982)	1000 i.u.D_4-D_{19}	$6/_{18}$(33)	$9/_{22}$(41)	+8
Santez-Valadez et al.(1982)	5000 i.u.D_{15}	$64/_{114}$(56)	$76/_{114}$(67)	+11
Sreenan &Diskin (1983)	1500 i.u.D_{10}-D_{20}	$13/_{25}$ (52)	$17/_{26}$ (65)	+13
	1500 i.u.D_{10}-D_{35}	$9/_{20}$ (45)	$9/_{16}$ (56)	+11
	1500 i.u.D_{10}-D_{20}	$19/_{28}$ (68)	$13/_{23}$ (57)	-11
Overall		$217/_{374}$	$220/_{341}$	
%		58.0	64.5	(P>0.05)

        While there was an improvement in conception rate this was
not significant within any one or all studies combined.
        From an examination of the literature and from the
experiments reported above the specific relationship between
progesterone on embryo survival remains unclear.   Progesterone
levels in the cycle preceeding insemination do not seem to
influence subsequent conception rate.   A higher progesterone
level in pregnant cows is evident from about Day 13 onwards
and it is likely that this increase is the result of a
luteotrophic signal from the embryo.
        Progesterone supplementation of subfertile cows has

resulted in some increase in pregnancy rates though the data reported here for dairy cows of normal fertility does not show any advantage in the use of exogenous progesterone in such animals. Likewise the use of hCG has not enhanced pregnancy rates. The administration of luteotrophic agents are unlikely to produce conclusive results until there is a better understanding of the endocrine process involved in luteolysis and the establishment of pregnancy.

156

REFERENCES

Ayalon, N. (1978). A review of embryonic mortality in cattle. J. Reprod. Fert. 54, 483-493.

Beal, W.E., Lukaszewska, J. & Hansel, H. (1981). Luteotrophic effects of bovine blastocysts. J. Anim. Sci. 52, (3) 567-574.

Betteridge, K.J., Eaglesome, M.D., Randall, G.C.B., Mitchell, D. & Sugden, E.A. (1978). Material progesterone levels as evidence luteotrophic or antiluteolytic effects of embryos transferred to heifers 12-17 days after oestrus. Theriogenology, 9, 86.

Betteridge, K.J., Randall, G.C.B., Eaglesome, M.D. & Sugden, E.A. (1984). The influence on pregnancy of PGF_2 secretion in cattle. 1. Concentrations of 15-Keton 13, 14^2dihydroprostaglandin F_2 and progesterone in peripheral blood of recipients of transferred embryos. Animal Reprod. Sci. 7,195-216.

Bishop, M.W.H. (1964). Paternal contribution to embryonic death. J. Reprod. Fert. 7,383-396.

Boyd, J. & Reed, H.C.B. (1961). Investigations into the incidence and causes of infertility in daisy cattle. Influence of some management factors affecting the semen and insemination conditions Brit. Vet. J. 117,74-86.

Bulman, D.C. & Lamming, G.E. (1978). Milk progesterone levels in relation to conception, repeat breeding and factors influencing acyclicity in dairy cows. J. Reprod. Fert. 54,447-458.

Christie, W.B., Newcomb, R. & Rowson, L.E.A. (1979). Embryo survival in heifers after transfer of an egg to the uterine horn contra-lateral to the corpus luteum and the effect of treatment with progesterone or HCG on pregnancy rates. J. Reprod. Fert. 56, 701-706.

Corah, L.R., Quealy, A.P., Dunn, T.G., & Kallenback C.C. (1974). Pre-partum and post-partum level of progesterone and Oestradiol in beef heifers fed two levels of energy. J. Anim. Sci. 39,(2) 380-385.

Dawson, F.L.M. (1954). Progesterone in functional infertility of cattle. Vet. Rec. 66,(23) 324-326.

Diskin, M.G. & Sreenan, J.M. (1980). Fertilization and embryonic mortality rates in beef heifers after artificial insemination. J. Reprod. Fert. 59, 463-468.

Erb, R.E., Garverick, H.A., andel, R.D., Brown, B.L., Callahan, C.J. (1976). Profiles of reproductive hormones associated with fertile and non-fertile inseminations of dairy cows. Theriogenology, 5, 227-242.

Folman, Y., Rosenberg, M. Herz, Z. & Davidson (1973). The relationship between plasma progesterone concentration and conception in post-partum dairy cows maintained on two levels of nutrition. J. Reprod. Fert. 34,267-278.

Gayerie, F. (1982). The application of cytogenetic to embryo mortality. Proc. SSAB Summer meeting, University of Nottingham School of Agriculture, Sutton Bonnington.

Greve, T. & Lehn-Jensen, H. (1982). The effect of HCG administration on pregnancy rate following non-surgical transfer of viable bovine embryos. Theriogenology 17,91.

Hasler, J.F., Bowen, R.A., Nelson, L.D. & Seidel, G.E. Serum progesterone concentrations in cows receiving embryo transfers. J. Reprod. Fert. 58, 71-78.

Henrick, D.M., Dickey, J.F., Niswender, C.D. (1970). Serum luteinizing hormone and plasma progesterone levels during the oestrous cycle and early pregnancy in cows. Biol Reprod. 2, 346-351.

Henricks, D.M., Lamond, D.R., Hill, J.R., Dickey, J.F. (1971). Plasma progesterone concentrations before mating and in early pregnancy in the beef heifer. J. Anim. Sci., 33, 450-454.

Herrick, J.B. (1953). Clinical observation of progesterone therapy in repeat breeding heifers. Vet. Med. 489-490.

Holness, D.H., Ellison, J.A., Willuns, L.M. (1977). Conception of beef cows in relation to concentration of progesterone in peripheral blood. Anim. Breed. Abs. 46, (5) 2070.

Holness, D.H., Ellison, J.A., Sprowson, G.W., Carvalho, A. de. (1977) Aspects of fertility in Friesland dairy cows with particular reference to the concentrations of progesterone in peripheral plasma Rhodesian J. Agric. Res. 15, (2) 109-117.

Holness, D.H., McCabe, C.T. & Sprowson, G.W. (1982). Observations on the use of human chorionic gonadotrophin (HCG) during the post-insemination period on conception rates in synchronised beef cows with subsoptimum reproductive performances. Theriogenology 17, (2) 133-140.

Johnson, K.R., Ross, R.H., Fourt, D.L. (1958). Effect of progesterone administration on reproductive efficiency. J. Anim. Sci., 17, 386-390.

Jones, G.S. (1976). The luteal phase defect. Fert. and Ster. 27, 351-356.

Jones, G.S., Aksel, S. & Wentz, A.C. (1974). Serum progesterone values in the luteal phase defects. Effect of chorionic gonadotrophin. Obstet. Gynecol. 44, 26.

Kunkel, R.N., & Stricklin, W.R. (1978). Donor-Recipient asynchrony, stage of embryo development and post-transfer survival of bovine embryos. Theriogenology 9, (1) 96.

Lawson, R.A.S. & Cahill, L.P. (1983). Modification of the embryo-maternal relationship in ewes by progesterone treatment early in the oestrous cycle. J. Reprod. Fert., 67, 473-475.

Lawson, R.A.S., Parr, R.A. & Cahill, L.P. (1983). Evidence of maternal control of blastocyst growth after asynchronous transfer of embryos to the uterus of the ewe. J. Reprod. Fert. 67, 477-483.

Linares, T. Larsson, K. & Edquivist, L.E. (1982). Plasma progesterone levels from oestrous through to Day 7 after A.I. in heifers carrying embryos with normal or deviating morphology. Theriogenology 17, (2) 125-132.

Lukaszewska, J. & Hansel, W. (1980). Corpus luteum maintenance during early pregnancy in the cow. J. Reprod. Fert. 59, 485-493.

Maurer, R.R. & Echternkamp, S.E. (1982). Hormonal asynchrony and embryonic development. Theriogenology 17, (1) 11-22.

Northley, D.L., French, L.R. (1978). Effect of embryo removal on bovine inter-oestrus interval. J. Anim. Sci. 47, Suppl. 1) 380.

Pope, G.S., Gupta, S.K. & Munro, I.B. (1969). Progesterone levels in the systemic plasma of pregnant cycling and ovariectomized cows. J. Reprod. Fert. 20, 369-381.

Randel, R.D., Garverick, H.A., Surve, A.H., Erb, R.E. & Callahan, C.J. (1971). Reproductive steroid in the bovine V comparisons of fertile and non-fertile cows 0-42 days after breeding. J. Anim. Sci., 33, 104-114.

Robertson, H.A. & Sarda, I.R. (1971). A very early pregnancy test for mammals; it's application to the cow, ewe and sow. J. Endocrinology 49, 407-419.

Roche, J.F., Boland, M.P. & McGeady, T.A. (1981). Reproductive wastage following artificial insemination of heifers. Vet. Rec. 109, 401-403.

Rosenberg, M., Gerz, Z., Davidson, M. & Folman, Y. (1977). Seasonal variations in postpartum plasma progesterone levels and conception in primiparous and multiparous dairy cows. J. Reprod. Fert. 51, 363-367.

Ross, G.T. (1979). Human chorionic gonadotrophin and maternal recognition of pregnancy. Cibe-Foundation Symposium 64, 191-201.

Rowdon, L.E.A., Moor, R.M. & Lawson, R.A.S. (1969). Fertility following egg transfer in the cow; effect of method, medium and synchronisation of oestrus. J. Reprod. Fert. 18, 517-523.

Thompson, F.N., Clekis, T., Kiser, T.E., Chen, H.J. & Smith, C. (1980). Serum progesterone concentrations in pregnant and non-pregnant heifers and after gonadotrophin releasing hormone in luteal phase heifers. Theriogenology 13, 407-417.

Santos-Valadez, S. de, Seidel, G.E. & Elsden R.P. (1982). Effect of HCG on pregnancy rates in bovine embryo transfer recipients. Theriogenology 17, 85.

Shemesh, M. (1980). Progesterone cyclicity and the effect of the conceptus or plasma progesterone in cattle and sheep. Proc. Int. Congr. Anim. Reprod. Artif. Insemin. 2, 103-108.

Shemesh, M., Ayalon, N. & Lindner, H.R. (1968). Early effects of conceptus on plasma progesterone level in the cow. J. Reprod. Fert. 15, 161-164.

Spitzer, J.C., Niswender, G.D., Seidel, G.E. Jr. & Wiltbank, J.N. (1978). Fertilisation and blood levels of progesterone and LH in beef heifers on a restricted energy diet. J. Anim. Sci., 46, 1071-1077.

Sreenan & Diskin, (1983). Early embryonic mortality in the cow; it's relationship with progesterone concentration. Vet. Rec. 112, 517-521.

Veenhuizen, E.L., Wagner, J.F. & Tonkinson, L.V. (1972). Corpus luteum response to 6-chloro- 17-aceto oxy-progesterone and HCG in the cow. Biol. Reprod. 6, 270-276.

Wagner, J.F., Veenhuizen, E.L., Tominson, L.V. & Rathmacher, R.P.(1973). Effect of placental gonadotrophin on pregnancy rate in the bovine. J. Anim. Sci. 36, 1129-1136.

Wilmut, I. & Sales, D.I. (1981). Effect of an asynchronous environment on embryonic development in sheep. J. Reprod. Fert. 61, 179-184.

Wiltbank, J.N., Rothlisberger, J.A. & Zimmerman, D.R. (1961). Effect of human chorionic gonadotrophin on maintenance of the corpus luteum and embryonic survival in the cow. J. Anim. Sci., 20, 827-829.

Wiltbank, J.N., Hawk, H.W., Kidder, H.E., Black, W.G., Ulberg, L.C., & Casida, L.E. (1956). Effect of progesterone therapy on embryo survival in cows of lowered fertility, J. Dairy Sci. 39, 456-461.

PREGNANCY DIAGNOSIS AND EMBRYONIC MORTALITY IN THE COW

V. Beghelli*, C. Boiti*, E. Parmigiani** and S. Barbacini**
*Istituto di Fisiologia Veterinaria
Universita di Perugia, Italy
**Istituto di Clinica Ostetrica e Ginecologia Veterinaria
Universita di Parma, Italy

ABSTRACT

A total of 3869 cows divided in five groups underwent three different methods of pregnancy diagnosis. Pregnancy was diagnosed in the first three groups by means of the progesterone milk test and in the other two groups by ultrasound and manual methods respectively. All three methods were suitable for evaluating either directly or indirectly the incidence of embryonic mortality (EM) and some of the limitations are discussed.

INTRODUCTION

Early pregnancy diagnosis (EPD) is one of the most important features of good reproductive management in dairy and beef farms. The techniques devised so far to obtain a reliable early pregnancy diagnosis are many: the early pregnancy factor (EPF) of foetal and maternal origin, detectable from about 24 to 48 hours after conception by means of the rosette inhibition method (Kock et al., 1983), the failure to return to oestrus at the expected time after insemination, the measurement of milk or plasma progesterone levels (Laing and Heap, 1971; Leopold and Seren, 1975; Erb et al., 1976; Gunzler et al., 1979; Beghelli et al., 1977a; Beghelli et al., 1977b), ultrasound techniques (Palmer and Driancourt, 1980; Reeves et al., 1984), and transrectal palpation of the uterus (Ball and Carroll, 1963; Ball and Parmigiani, 1980). All of these techniques have been shown to be efficient, but their reliability is different probably because of early and late embryonic mortality (EM).

The purposes of this study were to;

1. Confirm the efficiency of progesterone level (in milk or plasma), ultrasound techniques, and manual methods for assessing early pregnancy diagnosis (EPD) in the cow.

2. Evaluate the supposed embryonic mortality by using these three techniques.

MATERIALS AND METHODS

The experiments were conducted during the last three years on 3869 cows divided into five groups. Group A (Table 1) consisted of 2975 cows in which progesterone level was measured in a single sample collected 21 to 23 days after insemination. Group B (Table 2) was composed of 247 cows in which progesterone was measured at two periods; the first at insemination and the second 21 to 23 days later. This allowed confirmation of oestrus at the time of insemination. Group C (Table 3) was composed of 60 dairy cows in which milk samples were collected every other day for about four months postpartum and were assayed for progesterone. Milk samples were assayed for progesterone by a rapid direct radioimmuno-assay without previous solvent extraction (Simontacchi and Boiti, 1980; Boiti et al., 1974). Positive, negative or doubtful diagnoses were evaluated by comparison with two pools of milk obtained from pregnant and non-pregnant cows. Progesterone levels $\geqslant 4$ ng/ml were considered as positive EPD while progesterone values $\leqslant 2$ ng/ml as negative EPD. EPD was considered as doubtful when the porgesterone concentration was between the two discriminatory values.

For the heat detection test, milk progesterone levels lower than 2.0 ng/ml were considered as normal, and confirmatory of true oestrus, while progesterone levels greater than 2.0 ng/ml were considered abnormal. All the milk samples collected 21 to 23 days after AI were obtained from cows not returning to oestrus. Therefore, the rate of late EM was assessed in group A by high progesterone levels 21 to 23 days after AI together with a return to oestrus after 24 days or with a negative transrectal palpation. In group B, EM could be assessed more precisely because the only cows taken into account were those inseminated at the appropriate time. In Group C, also the lifespan of the corpora lutea subsequent to embryonic mortality has been analyzed in more detail. Group D, (Table 4) consisted of 22 cows and pregnancy diagnosis was performed by means of ultrasound techniques between 27 and 29 days of gestation. The ultrasound equipment employed was a

"Toshiba" real time ecograph model SAL 32A linear scanner with
an impermeable transrectal probe capable of producing and
receiving ultrasound at a frequency range of 39 megahertz.
The diagnoses and the eventual true embryonic mortality (EM)
were checked three months later by manual methods.   Group E,
(Table 4) was composed of 565 animals ranging from 33 to 45
days of gestation.   In this group, pregnancy was assessed
by manual methods as following: 1) Palpation of the uterus
for fluctuation of fluid (F) in 330 animals;   2) Palpation for
fluctuation, then for the amniotic vesicle (F+A) in 123
subjects;   3) Palpation for fluctuation and for slip of
the corioallantois membranes (F+M) in 112 cows.   In addition
these diagnoses and eventual embryonic mortality have been
checked within three months.

RESULTS
     Hormonal and clinical diagnoses were available for 2975
cows and results are presented in Table 1.

Table 1:   Group A. Accuracy of the progesterone test (EPD)
according to clinical diagnosis performed in 2975 cows 40 to
60 days after artificial insemination

	POSITIVE EPD	NEGATIVE EPD	DOUBTFUL EPD
Number of cows	2045	705	225
% of total cows	68.7	23.7	7.6
Pregnant cows	1327	28	9
% of accuracy	64.9	96.0	96.0

     The accuracy of positive EPD evaluated by clinical
diagnosis performed 45 to 60 days after insemination was
64.9% (1327/2045).   For the negative diagnoses, the EPD
test showed an accuracy of 96%, and similar accuracy (96%)
was found regarding doubtful EPD.
     Group B (Table 2) shows that the positive EPD were 185
(74.8%), the negative 44 (17.8%), and the doubtful 18 (7.3%).
The positive EPD evaluated by the clinical diagnoses were 116

(62.7%), the negative 2 (94.4%), and the doubtful 1 (94.4%).
In this group 69 cows with positive EPD were then diagnosed
non pregnant by the veterinarian.   However, we must point
out on the basis of progesterone measurement performed at
oestrus that 30 out of the 69 cows were not inseminated at the
correct time, therefore only 39 cases (21%), are imputable
to late embryonic mortality.

Table 2: Group B. Accuracy of the progesterone test (EPD)
according to clinical diagnosis

	POSITIVE EPD	NEGATIVE EPD	DOUBTFUL EPD
Number of cows	185	44	18
% of total cows	74.8	17.8	73
Pregnant cows	116	2	1
% of accuracy	62.7	95.4	94.4

The results obtained from the 60 dairy cows in group C are
summarized in Table 3.   The number of cows with positive EPD
was 45(75%), the negative EPD were 14 (23.3%), and only one
(1.7%) doubtful.   The accuracy of progesterone test compared
with manual diagnosis performed 45 to 60 days after AI has
been 73.3%, and 100% both for negative and doubtful diagnoses.
Twelve cows of this group with positive EPD failed to conceive.
According to progesterone profiles, the evaluation of these 12
unsuccessful AI could be explained as follows: EM in 8 cases
(17.7%); AI performed during luteal phases in 3 cows (6.6%);
while the last cow exhibited a short cycle after insemination.

Table 3:   Group C.   Accuracy of the progesterone test (EPD)
according to clinical diagnosis in 60 dairy cows

	POSITIVE EPD	NEGATIVE EPD	DOUBTFUL EPD
No. of cows	45	14	1
% of total cows	75.0	23.3	1.7
Pregnant cows	33	-	-
% of accuracy	73.3	100.0	100.0

In group D (Table 4), out of 12 pregnancy diagnoses made by
ultrasound techniques, 11 (92.5%) were manually confirmed
pregnant 3 months later, and only one case (7.5%) represented
true EM.   All the 10 negative diagnoses were confirmed correct
within 3 months.

In Plate 1 is presented an ultrasound image of a 29 day
pregnancy.

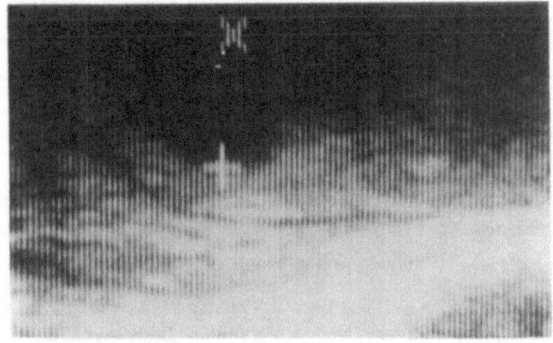

Plate 1.   Ultrasound image of a conceptus at 29 days of
pregnancy in a cow.

Table 4.   Group D.   Accuracy of pregnancy diagnosis by means
of real-time ultrasound technique performed at 27 to 29 days
of gestation

	POSITIVE	NEGATIVE
Ultrasound diagnosis	12	10
Manual diagnosis	11	10
% of accuracy	92.5	100

In group E (Table 5) 6 (1.81%) of 330 cows diagnosed
pregnant by palpation of the uterus for fluctuation of fluid,
were found to be non pregnant subsequently; 5 (4.07%) of 123
cows diagnosed pregnant by palpation for fluctuation and for
amniotic vesicle were subsequently non-pregnant, while 5
(4.46%) of 112 cows diagnosed pregnant by palpation for
fluctuation and slip of the choriollantois membrane were found
to be non pregnant subsequently.   In total, 16 (2.82%) of
565 cows diagnosed pregnant by the three methods of palpation
have presented EM.

Table 5:   Group E. Pregnancy diagnosis by manual palpation
performed at 33 to 45 days of gestation and embryonic mortality

PALPATION METHODS	No. of cows	No. of EM (%)
Fluctuance	330	6 (1.8)
Fluctuance & Amnios	123	5 (4.07)
Fluctuance & Slipping	112	5 (4.46)
Total	565	16 (2.82)

DISCUSSION AND CONCLUSIONS

     Early and late embryonic mortality (EM) directly reduces
reproductive efficiency in the cow causing substantial losses
in animal production.   The true incidence of EM however,
can not be accurately estimated because a reliable non-
surgical method for early pregnancy diagnosis has not yet been
devised.   The factors causing embryonic mortality are not
as yet clearly defined (Sreenan & Diskin, 1983).   Furthermore,
it is particularly difficult to state the rate and exact time
when embryonic mortality occurs, because embryonic losses
happening about Day 16 of the cycle do not alter either the
normal length of the oestrous cycle or the progesterone
concentration in milk or plasma at Days 21-23 (Northey &
French, 1980).   In the field, the earliest indication of
pregnancy in cattle is the failure to return to oestrus at
the expected time after insemination.   This parameter, however,
probably overestimates the rate of embryonic mortality by a
variable factor directly correlated with the heat detection
efficiency in any individual herd or by anoestrous cows
(Humblot, 1982)

     After the first suggestion that progesterone measurement
in milk or plasma could be used for the determination of early
pregnancy diagnosis (EPD) in cows (Heap et al., 1973) the
progesterone RIA or EIA technique has been widely employed in
many countries not only for EPD but also for confirmation of
oestrus (Gunzler et al., 1979), ovulation timing and ovarian
function (Erb et al., 1976).   Also late EM can be
conveniently evaluated 21 to 23 days after insemination by

the EPD test or, more accurately, by profiling progesterone in frequently collected milk samples. Our results from AI point out that the progesterone test performed 21 to 23 days after AI or mating, has an accuracy of about 65% and a supposed late EM of 35%. This percentage probably overestimates true embryonic mortality because false positive diagnoses from cows with "silent heats" two or three days before or after the sampling day are not taken into account. Also cows inseminated during the luteal phase, or with ovarian luteal cysts, or with a pathological condition of the uterus are not taken into account. This error can be avoided or substantially reduced by evaluating progesterone levels at the time of insemination and 21 to 23 days later as demonstrated by the results shown in Table 2 where out of 69 cows with positive EPD, 30 were inseminated at an incorrect time. So the supposed EM falls to 21%. The results of Table 3 confirm this finding and allow also to define the percentage of embryonic mortality and other causes of incorrect positive EPD.

The ultrasonic device is probably the most reliable technique for stating embryonic mortality, because it allows us to look at the embryonic vesicle, follow it's development and it's eventual disappearance. The only inconvenience is that ultrasound pregnancy diagnosis can not be performed before Day 27 of gestation. This situation could improve if a more sophisticated probe capable of visualizing an earlier embryonic vesicle were developed. The traditional manual methods of diagnosis have shown that embryonic mortality accounts for an average of 2.8% after 33 days of gestation. Besides this, our results suggest that pregnancy diagnosis by manual methods should be performed preferable by palpation of the fluctuation of the pregnancy uterine horn instead of palpating amniotic vesicle or slipping fetal membranes. Our overall results suggest also that the most consistent rate of EM occurs probably between Days 21 to 33 of the gestation period.

In conclusion we can say that:

1.   Measurement of progesterone level in milk or in plasma is an efficient method of indirect pregnancy diagnosis, especially if it is based on two samples, collected first at insemination and again 21 to 23 days later.   Both ultrasound and manual diagnoses are reliable direct means of pregnancy diagnosis after 27 and 33 days of gestation respectively.

2.   The apparent rate of embryonic mortality after Day 21 is higher when EPD is based on the progesterone test than when it is based on either ultrasound or manual methods. This arises partly because some cows may not have been in oestrus at the time of insemination and partly because progesterone level may remain high even though the embryo is already dead from about Day 16.

REFERENCES

Ball, L. and Carroll, E.J. 1963.   Induction of fetal death in cattle by manual rupture of the amniotic vesicle. JAVMA, 142, 373-374.

Ball, L. and Parmigiani, E. 1980.   Nuovi aspetti della diagnosi di gravidanza mediante esplorazione rettale nella bovina.   La Clinica Veterinaria, 103. 6, 309-316.

Beghelli, V., Boiti, C., Ceccarelli, P. and Debenedetti, A. 1977a.   Dosaggio del progesterone e sue pratiche applicazioni in campo zootecnico.   Nota 1: Valutazione della funzionalita ovarica. Atti Soc. It. Sci. Vet. 31, 432-434.

Beghelli, V., Boiti, C., Ceccarelli, P. and Debenedetti, A. 1977b.   Dosaggio del progesterone e sue pratiche applicazioni in campo zootecnico.   Nota 11: Diagnosi di gravidanza precoce. Atti Soc. It. Sci. Vet., 31, 435-437.

Boiti, C., Ceccarelli, P., Olivieri, O., Daniotti, P. and Pennis, F. 1974. Messa a punto di un metodo per la determinazione radioimmunogogica del progesterone plasmatico. Atti Soc. It. Sci. Vet., 28, 366-371.

Erb, R.E., Garverick, H.A., Randel, R.D., Brown, B.L. and Callahan, C.J. 1976.   Profiles of reproductive hormone associated with fertile and nonfertile inseminations of dairy cows.   Theriogenology, 5, 227-242.

Gunzler, O., Rattenberger, E., Gorlach, A., Ganh, R., Hocke, P., Claus, R. and Karg, H. 1979.   Milk progesterone determination as applied to the confirmation of oestrus, the detection of cycling and as an aid to veterinarian and biotechnical measures in the cows.   Br. Vet. J., 135, 541-549.

Heap, R.B., Gwyn, M. Laing, I.A. and Walters, D.E. 1973.

Pregnancy diagnosis in cows. Changes in milk progesterone concentration during the oestrus cycle and pregnancy
measured by a rapid radioimmunoassay. J. Agric. Sci.
Camb., 81, 151-157.

Humblot, P. 1982. Respective incidence of late embryonic
mortality and post insemination anoestrus in late
returns to oestrus in dairy cows. In "Factors
influencing fertility in the poaspartum cow" (Eds. H.
Karg and E, Schallenberg). (Martinus Nijhoff Publsrs.)
pp. 298-304.

Laing, J.A. and Heap, R.B. 1971. The concentration of progesterone in the milk of cows during the reproductive
cycle. Br. Vet. J., 127, XIX.

Leopold, A. and Seren, E., 1975. Indagine sull'incidenza
della mortalita embrionale nella bovina di razza
frisona. Atti Soc. It. Sci. Vet., 29, 354-361.

Koch, E., Morton, H. and Ellendorff, F. 1983. Early
pregnancy factor: biology and practical application
Br. Vet. J., 139, 52-58.

Northey, D.L. and French, L.R. 1980. Effect of embryo removal
and intaruterine infusion of embryonic homogenates on
the lifespan of the bovine corpus luteum. J. Anim.
Sci., 50, 298-302.

Palmer, E. and Friancourt, M.A. 1980. Use of ultrasonic
echography in equine ginecology. Theriogenology, 13,
203-207.

Reeves, J.J., Rantanen, N.W. and Hauser, M. 1984. Transrectal real-time ultrasound scanning of the cow
reproductive tract. Theriogenology, 21, 485-494.

Simontacchi, C. and Boiti, C. 1980. Metodo rapido per il
dosaggio del progesterone e sua applicazione nella
pratica clinica. Atti. Soc. It. Buiatria, 12, 1-7.

Sreenan, J. M. and Diskin, M.G. 1983. Early embryonic
mortality in the cow: It's relationship with progesterone
concentration. The Vet. Rec., 28, 517-521.

ACKNOWLEDGEMENT

This research was supported in part by grant of
Consiglio Nazionale della Ricerche, P.F. IPRA contratto n.
83.01505.55, in part by fund of Regione Umbria, Assessorato
Agricoltura with the co-operation of Breeding Associations of
Perugia and Terni.

Appreciation is expressed to Mr. L. Mariucci and C.
Canali for technical assistance.

DISCUSSION

Bolet:

What is the repeatability of progesterone level in sheep?

Wilmut:

Approximately 15% for baseline and luteal levels suggesting little opportunity for genetic selection for increased progesterone level.

Greve:

Did you examine the relationship between progesterone and oestrogen levels around the time of oestrus with subsequent ovulation rate?

Wilmut:

We have only defined the correlation between luteal phase progesterone and ovulation rate.

Heap:

Is there a significant rise in progesterone concentration during the baseline period.   If so, you should compare this with embryo survival rather than calculating a mean value for the baseline period?

Wilmut:

We will have to examine this possibility further but my impression is that progesterone level does not increase generally.

Heap:

You have interpreted the significant correlation between baseline progesterone and embryo survival in terms of the effect on uterine environment.   Can you exclude an effect on the embryo during tubal life?

Wilmut:

No.   The effect may also arise at the oviducal stage.

Boland:

Were the Day 3 embryos transferred to the uterus and was the lower progesterone level in the out-of season ewes due to a lower ovulation rate or to inadequate C.L. function?

Wilmut:

Yes and progesterone profiles were influenced both by ovulation rate and season.

Greve:

What effect had exogenous progesterone on circulating levels?

Diskin:

Progesterone supplementation did not affect circulating progesterone level at either 6 or 24 hours after administration The exogneous progesterone decreased C.L. weight and probably endogenous progesterone production.

Heap:

Is there evidence for improved fertility in cows implanted with progesterone releasing devices shortly after breeding?

Diskin:

Not substantial evidence. Such devices (PRIDS) release progesterone rapidly at first which then falls to ineffective levels. There is also evidence that the admininstration of high levels of progesterone early in the cycle can cause premature C.L. regression and short cycles.

Niemann:

What is the optimum progesterone level for pregnancy maintenance and is progesterone supplementation useful in embryo transfers?

Diskin:

There is no optimum level. We have recorded plasma

progesterone levels as low as 2 ng per ml in pregnant controls
and as high as 45 ng per ml in pregnant hCG treated animals.
There is no evidence for a beneficial effect of exogenous
progesterone in recipients.

Fischer:

Progesterone supplementation increased the pregnancy rate
in heifers by 22%, would you comment on this?

Diskin:

Fertility in heifers is lower than in cows.   Perhaps
this arises from a proportion of heifers having inadequate
luteal function and that the exogenous progesterone
overcame this.

Boland:

At what stage was pregnancy diagnosis by palpation
carried out?

Parmigiani:

From Day 33 to 45 after breeding.

Sreenan:

Can you detect multiple pregnancy in the cow by
ultrasound?

Parmigiani:

No but I understand that there is evidence in the
literature that mulptiple pregnancy can be detected in mares
as early as about Day 17 so this may be an indication that is
is possible in the cow.

EMBRYONIC MORTALITY IN STEROID IMMUNIZED SHEEP

M.P. Boland[*], J.D. Murray, R.J. Scaramuzzi, C.D. Nancarrow,

R.M. Hoskinson, R. Sutton, I. Hazelton

CSIRO Division of Animal Production
P.O. Box 239, Blacktown, N.S.W., Australia 2148
*University College, Dublin, Ireland

ABSTRACT

One hundred and seventy six mature Merino ewes were used in an experiment to study the effects of immunization against androstenedione on fertilization rates and early embryonic mortality. Animals were immunized such that the interval from booster injection to mating was controlled to give either high or low antibody titre levels. Immunization significantly increased ovulation rates, being 1.49, 2.23 and 1.88 for control, high and low groups respectively. There was a significantly lower mating response in the high titre group. At day 2 post mating, egg recovery rate was decreased by immunization ($P<0.05$), and fertilization rate was higher in controls than in low titre group ($P<0.05$). At day 13 there was no difference in recovery rates or in the proportion of ewes which were pregnant. At days 25-29 there was a further decrease in pregnancy rate in all groups, but the difference between the groups was not significant.

Chromosome abnormalities were found in 11% of day 2 embryos, with no differences between treatment groups. At day 13 there was evidence of chromosome abnormality in less than 2% of blastocysts, while there was no evidence of abnormalities at days 25-29. Results of the present study indicate that embryo loss is high in androstenedione immune Merino ewes, particularly if the interval from booster injection until mating is too short.

INTRODUCTION

Over half a century has elapsed since reports of Hammond, (1921) and

Corner, (1923) alluded to the possible loss of reproductive potential in

farm animals caused by the failure of fertilized eggs to develop normally

to gestation and parturition. It is now well established that lambing

rates in ewes are generally lower than fertilization rates, especially

when multiple ovulation has occurred. This loss which is of serious

economic consequences has been attributed to embryonic mortality. Numerous attempts have been made to define this loss which is known to occur at

several stages between fertilization and lambing (Edey, 1969). A variety

of experimental treatments has been applied in the study of embryonic mortality in sheep, including nutritional, seasonal and progestational treatments (Edey, 1976; Kelly and Allison, 1976; Parr, Cumming and Clarke, 1982;

Lunstra and Christenson, 1982). Embryo survival tends to be higher in

single as opposed to twin ovulating ewes (White, Rizzoli and Cumming, 1981).

In recent years there has been a growing demand for increased output from the sheep flock. Prolificacy may be improved genetically by crossing and environmentally by improved nutrition at or around the time of mating. Recently the techniques of active immunization against ovarian steroids have been shown to alter the feedback balance in favour of additional ovarian activity. Immunization against androstenedione, oestradiol and oestrone and testosterone have all been shown to cause increased follicular growth (Scaramuzzi, Davidson and Van Look, 1977; Scaramuzzi; Martensz and Van Look, 1980; Scaramuzzi, Baird, Martensz, Turnbull and Van Look, 1981). A commercial preparation (Fecundin, Glaxo Pty, Australia) for immunization against androstenedione has been launched on the market in Australia and New Zealand. The presence of antibodies directed against androstenedione leads to a presumed reduction in the levels of biologically available androstenedione and increases in ovulation rates (Scaramuzzi et al., 1977). Large scale farm trials, using Fecundin, have been shown to increase litter sizes in certain breeds and crosses (Geldard, Scaramuzzi and Wilkins, 1983).

The present experiment was designed to study the level of fertilization failure and embryonic mortality in Merino ewes following immunization against androstenedione. In addition, the rate of embryo development was studied, together with the incidence of chromosome abnormalities in these embryos.

MATERIALS AND METHODS

One hundred and seventy six mature Merino ewes were used and were divided into 3 treatment groups as follows : (a) untreated controls; (b) immunized against androstenedione - 3 HSA (Fecundin, Glaxo Pty, Australia), such that the interval from booster injection until mating was approximately 14-days. This group was estimated to have high antibody titres and was designated the high group; (c) immunized as in group (b), but with an estimated interval from booster injection until mating of 25-days and this was termed the low antibody group.

Experimental Procedure

All immunized ewes were given a primary injection of Fecundin and this was followed 28 days later by a booster injection (Table 1). The mating and embryo recovery were facilitated by synchronizing the cycles with 60 mg MAP sponges (Upjohn Ltd) for 14 days and breeding at the second post treatment oestrus. Ewes were bred at the second oestrus to avoid

TABLE 1.   Experimental protocol

Day	Controls	High Titre	Low Titre
1	-	-	Primary immunization
11	-	Primary immunization	-
20	Sponges in	Sponges in	Sponges in
28	-	-	Booster immunization
34	Sponges out	Sponges out	Sponges out
39	-	Booster immunization	-
48	Join with fertile rams	Join with fertile rams	Join with fertile rams

any carry-over effects that progestagen might have on fertility.   Two relays of 12 rams per 100 ewes were exchanged twice daily at the morning and evening oestrus checks.

Embryo Recovery

Embryo recovery was attempted at three different times post mating, i.e. day 2, 13 and 25-29.   Ewes were checked for oestrus marks twice daily and those destined for flushing on day 2 were subjected to surgery between 0900 and 1100 hr. or between 1900 and 2100 hr. which was approximately 48 hr. after detection of oestrus.   The oviduct was flushed with 5 ml. phosphate buffered saline (PBS) containing 5% heat-treated sheep serum (SS). At day 13 all animals were slaughtered and each horn was flushed with 15 ml. PBS.   The remaining ewes were joined with sterile teaser rams and those not returning to service were slaughtered between days 25 to 29.   The ovulation rate of ewes assigned to the 25 to 29 day group was assessed by endoscopic methods at days 7-12 of the cycle.   Egg recovery and fertilization rates were subjected to χ^2 analysis while ovulation rates were compared using analysis of variance.

Chromosome Methods

Embryos were cultured in PBS + 15% SS, 0.8 ng/ml colchicine, 100 i.u./ ml penicillin G and 100 ng/ml streptomycin sulphate, at 37°C in a 5% CO_2 atmosphere.   Embryos were cultured for varying time intervals according to age (Boland et al., 1984).

RESULTS

A lower proportion of high titre immunized ewes was bred in the 10 days following joining with fertile rams (P<0.05), and there were significant differences (P<0.01) in ovulation rates between all treatment groups (Table 2).

TABLE 2   Oestrous and ovarian response in control and androstenedione immune ewes

| | Group | | |
	Control	High	Low
No.ewes treated	59	58	59
No. (%) mated	57(97)	47(81)	56(95)
Mean ovulation rate	1.49	2.23	1.88
Mean no. large follicles	0.11	0.66	0.38

Egg recovery rates at day 2 were affected by treatment (Table 3).   A lower proportion (P<0.025) of eggs was recovered from the high titre group than from controls, and there was evidence of a lower fertilization rate in the low titre group when compared with controls (P<0.05).   There was no difference in the proportion of ewes yielding eggs or yielding fertilized eggs.

By day 13 there was no significant difference between the groups(Table 4), although a lower recovery rate and lower proportion of pregnant ewes was evident in the immunized groups.   At days 25-29 there was a further decrease in pregnancy rate in all groups, but the differences were not significant because of the small numbers involved (Table 5).   The number of embryos per pregnant ewe was slightly higher in both immunized groups.

The chromosome complement of day 2 embryos was determined and there were no significant differences between controls and immunized groups (Table 6).   The sex ratio did not vary from 1:1.   The overall incidence of chromosomally abnormal embryos was 11%.   By day 13 the level of chromosome abnormality had decreased to less than 2%, which suggests that those embryos which are likely to be lost due to chromosome abnormalities have disappeared. There was evidence that embryo length at day 13 and weight at days 25-29 was lower in immunized ewes compared with controls.

TABLE 3   Recovery and fertilization rates at day 2 post mating

	Group		
	Control	High	Low
No.of ewes examined	19	20	20
No. of ovulations	33	48	36
No.(%) eggs recovered	28(85)	28(58)	25(69)
No.(%) eggs fertilized	25(89)	20(71)	17(68)
No.(%) ewes yielding eggs	18(95)	17(85)	16(80)
No.(%) ewes yielding fer-tilized eggs	16(89)	14(82)	12(80)

TABLE 4   Recovery of blastocysts at day 13 in control and immunized ewes

	Group		
	Control	High	Low
No. of ewes examined	20	20	20
No.of ovulations	25	45	39
No.(%) eggs/embryos recovered	19(78)	30(67)	28(72)
No.(%) fertilized	18(95)	26(87)	27(96)
No.(%) normal blastocysts	18(100)	24(92)	23(85)
No.(%) ewes yielding eggs	17(85)	16(80)	16(80)
No.(%) ewes pregnant	16(80)	15(75)	14(70)
No.blastocysts/pregnant ewe	1.13	1.60	1.64

TABLE 5   Pregnancy rates at day 25-29 in control and immunized ewes

	Group		
	Control	High	Low
No. ewes mated	18	8	16
No. which repeated	4	5	5
No.(%) pregnant	13(72)	3(38)	10(63)
No. embryos/pregnant ewe	1.46	1.67	1.70
No. abnormal embryos	2	0	0

TABLE 6   Chromosome complements in day 2 Merino embryos (adapted from Murray et al., 1985)

Treatment Group	No. of Embryos examined	No. of Embryos yielding chromosomes	No. normal	(%) normal
Control	64	26	23	88.5
High	57	23	20	87.0
Low	56	24	22	91.7

DISCUSSION

Immunization of Merino ewes against androstenedione increased the ovulation rate as previously reported for crossbred ewes (Scaramuzzi et al., 1983). Fecundin treatment of Merino ewes may lead to problems of egg recovery and perhaps fertilization. Differences in recovery and fertilization rates have been observed in progestagen PMSG synchronized ewes (Lunstra and Christenson, 1981) but ewes in the present study were mated at a natural oestrus. The reduced recovery rates in the present study may indicate a possible failure of fimbrial collection, or a loss from the oviduct subsequent to ovulation. There was a difference in pregnancy rates of the order of 7-10% between days 2 and 13, which is comparable to the level of chromosomally abnormal embryos. By days 25-29 there was a further drop in pregnancy rate in all groups. There was no evidence of a drop in litter size in pregnant animals in any of the groups. The higher level of reproductive wastage in the immunized ewes could be partly due to the higher ovulation rate, as it has been suggested that embryo survival may be lower in twin ovulating ewes (White et al., 1981). Levels of embryonic mortality varying from 12 to 29% have been reported in control and progestagen synchro-

177

nized ewes (Lunstra and Christenson, 1981). The level of mortality appeared to be related to ovulation rate which was higher in the synchronized group (Lunstra and Christenson, 1981). Work by Parr et al., (1982) has shown that, despite live-weight losses from mating to days 11 and 21, ewes on 25% maintenance nutrition showed no evidence of increased embryonic mortality, although the embryos were lighter in ewes on the lower level of nutrition.

Many studies have involved slaughter of the ewes, and there is ample evidence that in some circumstances losses due to abnormalities in the fertilized eggs can be substantial (Edey, 1969). Data of Lunstra and Christenson, (1981) showed that increased embryonic mortality among treated ewes was associated with increased variation in the stage of embryo development within the ewe. Data from the present experiment would suggest that embryo development in the immunized ewes both at days 13 and 25-29 may be retarded relative to that in controls. Further work is necessary to elucidate factors responsible for this retarded development.

REFERENCES

Boland, M.P., Murray, J.D., Scaramuzzi, R.J., Moran, C., Sutton, R., Hoskinson, R.M., Hazelton, I. and Nancarrow, C.D. 1984. Reproductive wastage and early embryonic chromosomal abnormalities in immunized and control Merino ewes. In "Review of Research in Sheep Reproduction". Proc. Conf. Lorne, Aust. Acad. Technol. Sci. Canberra, Ed. D.R. Lindsay, pp. 137-139.
Corner, G.W. 1923. The problem of embryonic pathology in mammals with observations upon uterine mortality in the pig. Am. J. Anat., 31, 523-545.
Edey, T.N. 1969. Prenatal mortality in sheep: a review. Anim. Breed. Abstr., 37, 173-190.
Edey, T.N. 1976. Embryo mortality. In: Sheep Breeding: Proc. 1976 Int. Cong. Muresk and Perth. Eds. G.J. Tomes, D.E. Robertson and R.J. Lightfoot. pp. 400-410.
Geldard, H., Scaramuzzi, R.J. and Wilkins, J.F. 1984. Immunization against polyandroalbumin leads to increases in lambing and tailing percentages. New Zealand Vet. J., 32, 2-5.
Hammond, J. 1921. Further observations on the factors controlling fertility and foetal atrophy. J. Agric. Sci., Camb. 11, 337-366.
Kelly, R.W. and Allison, A.J. 1976. Returns to service, embryonic mortality and lambing performance of ewes with one or two ovulations. In: Sheep Breed: Proc. 1976 Int. Congr. Muresk and Perth. Eds. G. J. Tomes, D.E. Robertson and R.J. Lightfoot, pp. 418-423.
Lunstra, D.D. and Christenson, R.K. 1982. Fertilization and embryonic survival in ewes synchronized with exogenous hormones during the anoestrous and estrous seasons. J. Anim. Sci., 53, 458-463.

Murray, J.D., Boland, M.P., Moran, C., Sutton, R., Nancarrow, C.D., Scar-
    muzzi, R.J. and Hoskinson, R.M. 1985. The occurrence of haploid and
    haploid/diploid mosaic embryos in untreated and androstenedione immune
    Australian Merino sheep. J. Reprod. Fert. (In Press).
Parr, R.A., Cumming, I.A. and Clarke, I.J. 1982. Effects of maternal
    nutrition and plasma progesterone concentrations on survival and
    growth of the sheep embryo in early gestation. J. Agric. Sci., Camb.,
    98, 39-46.
Scaramuzzi, R.J., Davidson, W.G. and Van Look, P.F.A. 1977. Increasing
    the ovulation rate of sheep by active immunization against an ovarian
    steroid, androstenedione. Nature, Lond, 269, 817-818.
Scaramuzzi, R.J., Martensz, N.D. and Van Look, P.F.A. 1980. Ovarian mor-
    phology and the concentration of steroids and of gonadotrophins during
    the breeding season in ewes actively immunized against oestradiol-17β
    or oestrone. J. Reprod. Fert., 59, 303-310.
Scaramuzzi, R.J., Baird, D.T., Martensz, N.D., Turnbull, K.E. and Van Look,
    P.F.A. 1981. Ovarian function in the ewe after active immunization
    against testosterone. J. Reprod. Fert., 61, 1-9.
Scaramuzzi, R.J., Geldard, H., Beels, C.M., Hoskinson, R.M. and Cox, R.I.
    1983. Increased lambing percentages through immunization against
    steroid hormones. Wool Tech. and Sheep Breeding, 31, 87-97.
White, D.H., Rizzoli, D.J. and Cumming, I.A. 1981. Embryo survival in
    relation to number and site of ovulations in the ewe. Aust. J. Exp.
    Agric. Anim. Husb., 21, 32-38.

EMBRYO LOSS FOLLOWING EMBRYO TRANSFER IN CATTLE AND SWINE

Torben Greve and Marcelo Del Campo,*

Institute for Animal Reproduction,
The Royal Veterinary and Agricultural University,
13, Bulowsvej, DK-1870 Copenhagen V, Denmark

*Instituto Reproduccion Animal,
Facultad de Ciencias Veterinaris,
Universidad Austral del Chile, Chile

ABSTRACT

Embryo survival following transfer to synchronized recipients was analysed in the context of a review of literature. In the bovine species most embryos are transferred non-surgically giving rise to an average pregnancy rate of 58.8% (4638/7882) on day 60-90 following ipsilateral transfer of a single embryo to synchronized recipients. It is believed that interfarm variation and recipient quality contribute to the variation. The percentage of recipients exhibiting prolonged estrous cycles was 5-6%. The abortion rate following transfer has not received special attention, but may be a problem when transferring embryos from certain donors. To increase the twinning rate two embryos have been transferred bilaterally or unilaterally to synchronized recipients or one embryo has been deposited in the contralateral horn of already mated recipients. Of 590 bilaterally transferred embryos, 380 or 64.4% survived. The survival rate of embryos transferred to the contralateral horn could not be fully evaluated, but the survival rate of either of the embryos was 258 out of 425 (60.7%). Based on these reviews it can be concluded that the survival rate following the three different methods is almost identical (P, embryo survival = 0.59). Embryo survival and pregnancy rates following synchronous and asynchronous transfer in pigs has been reviewed. Pregnancy and embryonic survival rates were 75% and 71%, respectively, and synchronous transfer and negative synchrony allowed a higher survival than positive synchrony.

CATTLE

Development of reliable non-surgical transfer techniques has expanded the use of embryo transplantation in cattle considerably. Recent estimation has predicted that more than 100,000 calves will be born in 1985 following transfer of single or twin embryos. The embryo loss following transfer of fertilized 7-8 days old embryos is difficult to evaluate, because the available literature comprised a very heterogenous pool of data. The experiments have been conducted under different conditions with different recipient qualities, and frequently many operators have been implicated, and due to a limited number of eggs/transfers the material has been difficult to treat statistically. To make the following review more easily accessible the references will be divided into three

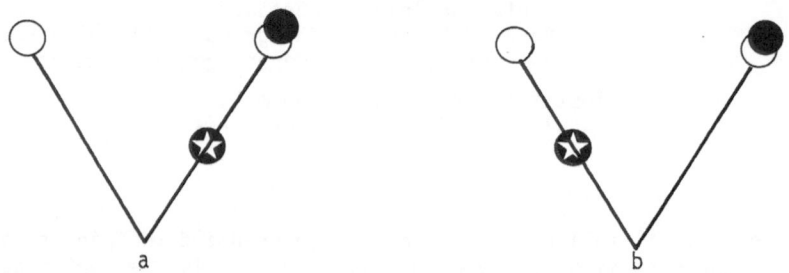

1. Single transfer to the ipsilatereal (a) or contralateral (b) horn

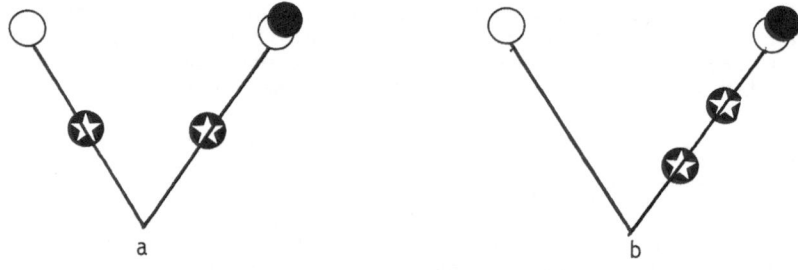

2. Twin transfer to each (a) or same (b) horn

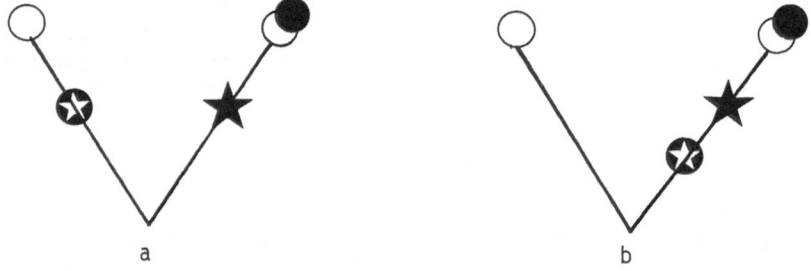

3. Single transfer to the contralateral (a) or ipsilateral (b) horn
of mated recipients

○ OVARY   ● CORPUS LUTEUM   ✪ TRANSF. EMBRYO   ★ INDIGENOUS EMBRYO

Fig. 1. Schematic presentation of different methods of transfer

categories:

1. Single embryo transfer i.e., transfer of one embryo to the ipsilateral (I) or the contralateral horn (C)

2. Twin transfer i.e., transfer of two embryos either both into one uterine horn or one embryo into each uterine horn

3. Transfer to previously inseminated recipients i.e., transfer of one embryo into the contralateral horn or the ipsilateral horn (PI)

The drawings in fig. 1 give a schematic presentation of the various methods of transfer. Within the single groups it is possible to perform either surgical (S) or non-surgical (N-S) transfer and to deposit the embryo at the base of the horn, in the middle part of the horn or in the tip of the horn. It is impossible to depict from each reference which positioning has been utilized.

1. SINGLE EMBRYO TRANSFER TO UNMATED RECIPIENTS (fig. 1.1.)

Although the initial experiments on non-surgical transfer (N-S) of bovine embryos were discouraging (Avery et al., 1962; Sugie, 1965; Rowson & Moor, 1966) it is now the predominant method of transfer. According to a recent survey made by the IETS (Embryo Transfer Newsletter 3, 1984) a very high proportion of all embryos are transferred by means of this simple N-S technique (Europe: 67% of 6293 embryos, Canada: 76% of 21,233 embryos and U.S.A.: 80% of 121,855 embryos). The pregnany rate varies from 19-65% as seen in Table 1, where ipsilateral transfer results, both surgical and non-surgical are presented. The more significant data are presented by Schneider et al. (1980) and Hahn (1984) who achieved pregnancy rates of 65% (1670/2556) and 59% (2348/3980), respectively, and the overall N-S pregnancy rate was found to be 58.8% (4638 out of 7882 recipients became pregnant). Results accumulated in Denmark from 1976 have given rise to a pregnancy rate of 51% (125 out of 246 recipients pregnant at 60 days).

Comparative non-surgical (N-S) and surgical (S) transfer studies have established that the latter method is superior (Newcomb, 1979; Newcomb & Rowson, 1980; Rowe et al., 1980). However, most of these experiments were based on a low number of embryos and it is doubtful whether there is any real difference when the N-S technique is done properly by a skilled operator.

Following transfer of day 7-10 embryos a certain proportion will succomb prior to 60-90 days. Thus, Bowen et al. (1978) found that 6% of all non-pregnant recipients at day 60 had increased cycle length (48-55 days) reflecting early embryonic loss. Markette et al. (1980) monitored plasma progesterone profiles of the recipients and noted that 6.1% of all recipients had cycle length prolongations (48-59 days) probably due to early embryonic mortality. Renard et al. (1980) studied the progressive embryonic loss following N-S transfer and found that the survival rate of non-cultured 10 day old bovine embryos decreased from 69.8% on day 21 to 49.3% on day 60. Cultured embryos experienced an even more drastic reduction in survival rate. The same trend in embryonic survival was revealed by Sreenan & McDonagh (1981).

An aspect that has not been discussed in great detail and of great importance is the abortion rate. Renard et al. (1980) reported that 4.4% aborted and Anderson et al. (1982) found that 2 recipients out of 26 (7.6%) aborted.

Recent results from our laboratory indicated that certain donor animals did contribute with a high incidence of abortions/resorption as seen in Table 2. Unfortunately, the aborted material was not suitable for chromosome analysis. It is known, however, that chromosomal abnormalities do occur in the embryos (Guyerie, 1983).

Several factors are known to influence the survival rate of ipsi-laterally transferred embryos. Besides culture time prior to transfer (Peters et al., 1978; Renard et al., 1980; Schneider et al., 1980), the ease of transfer (Boland, 1976), the skill of the operator (Bowen et al., 1978; Schneider et al., 1980), the instrumentation (Wilmut et al., 1978; Tervit et al., 1978; Jillella & Baker, 1977; Tervit et al., 1980; Takahashi et al., 1981), embryo quality (grade) and developmental stage and donor/recipient and embryo/recipient synchrony (Rowson et al., 1969; Schneider et al., 1980; Wright, 1981; Wright Jr. & Lindner, 1983). It is also believed that the quality of the recipient is important. A very decisive factor for embryonic survival is the site of transfer in relation to the ovary bearing the CL and contralateral deposition always results in a lower survival rate than ipsilateral transfer (Tervit et al., 1977; Christie et al., 1979; Newcomb & Rowson, 1980; Sreenan et al., 1981; Del Campo et al., 1983).

It has been suggested that lower survival rate of embryos transferred in the horn contralateral to the CL is due to the inhability of the embryos to prevent CL regression (Betteridge et al.,1976; Christie et al.,1979; Del Campo et al., 1983). The loss of contralateral embryos apparently is higher before day 30.

TABLE 1. Survey of results of non-surgical transfer of one embryo to the ipsilateral horn.

No. of transfers	Technique	Pregnant recipients	(%)	Reference	Year
14	N-S	0	(0)	Avery et al.	1962
83	N-S	17	(20)	Sugie	1965
14	N-S	3	(21)	Rowson & Moor	1966
18	S	13	(72)	Rowson et al.	1969
76 (day 14)	S	36	(58)	Betteridge et al.	1976
49	N-S	25	(52)	Alexander et al.	1977
48	N-S	17	(35)	Tervit et al.	1978
37	N-S	7	(19)	Wilmut et al.	1978
83	N-S	27	(32)	Brand et al.	1978
27	N-S	17	(63)	Hahn et al.	1978
48	N-S	22	(46)	Trounson et al.	1978
8	S	5	(63)	Peters et al.	1978
95	N-S	18	(19)	Bowen et al.	1978
36	S	20	(56)	Newcomb	1979
75	N-S	37	(49)	Newcomb	1979
32	N-S	18	(56)	Greve & Lehn-Jensen	1979
36	N-S	23	(64)	Nielsen & Østergård-Nielsen	1979
213	S	134	(63)	Markette et al.	1980
83	N-S (no culture)	41	(49)	Renard et al.	1980
45	N-S (culture)	15	(33)	Renard et al.	1980
68	S	42	(62)	Rowe et al.	1980
34	N-S	11	(33)	Rowe et al.	1980
50	N-S	31	(59)	Lampeter et al.	1980
72	N-S	31	(43)	Tervit et al.	1980
2256	N-S	1670	(65)	Schneider et al.	1980
19	S	6	(34)	Takahashi	1980
18	S	10	(56)	Takahashi	1980
22	N-S	13	(59)	Takahashi et al.	1981
119	N-S	62	(52)	Greve	1981
144	S	70	(49)	Greve	1981
22 Sheat +	N-S	14	(64)	Takahashi et al.	1982
24 Sheat -	N-S	8	(33)	Takahashi et al.	1982
3980	N-S	2348	(59)	Hahn	1984

N-S = Non-surgical transfer

S = Surgical transfer

TABLE 2. Resorption/abortion incidence in 7 Black and White and Red Danish donor animals.

Donor	N-S transfer	Pregnant at 6 weeks	Abortions
I	3	3	1 (10 weeks)
II	3	2	2 (8 months)
III	10	7	1 (60 days)
			1 (5 months)
			1 (5½ months)
IV	1	1	1 (8 months)
V	6	5	2 (55 days)
VI	5	3	2 (5 months)
VII	4	2	1 (4 months)

The table demonstrates that abortions may be a problem in certain donor cows.

2. TWIN EMBRYO TRANSFER

When 2 embryos are transferred for the production of twins they are inserted either both into one uterine horn (Rowson et al., 1971; Newcomb et al., 1980) or one in each horn (Sreenan and Beehan, 1976; Heyman and Renard, 1978; Newcomb et al., 1978; Anderson et al., 1978, 1982; Del Campo et al., 1983).

Unilateral twin embryo transfers (Fig.1, 2b)

The percentage of pregnancies early in gestation in unilateral tranfers has been found high (about 90%), but the percentage of twins born has been low (approx. 12%). The cause of the loss of pregnancies later in gestation has been attributed to the result of overcrowding in that horn (Rowson et al., 1971).

Bilateral transfers (Fig. 1, 2a)

In the case of bilateral transfers, a high pregnancy rate also has been reported, however the percentage of twin pregnancies has been quite inconsistent (see Table 3) and varies between 30-80%.

This is in agreement with naturally occurring multiple pregnancies (Cady and Van Vleck, 1978; Scanlon, 1972; Hanrahan, 1983).

Different laboratories (see Table 3) have been more concerned with improving the techniques reported earlier than in studying the cause of embryo loss.

The causes of embryo mortality in bilateral pregnancies have not been critically studied, but may involve reduced uterine capacity or a reduced embryonic mortality. Renard et al. (1980) studied the progressive embryonic loss following N-S transfer and found that the survival rate of non-cultured 10 day old bovine embryos decreased from 69.8 on day 21 to 49.3% on day 60. Cultured embryos experienced an even more drastic reduction in survival rate. The same trend in embryonic survival was revealed by Sreenan & McDonagh (1981).

An aspect that has not been discussed in great detail and that is of great importance is the abortion rate. Renard et al. (1980) reported that 4.4% of recipients aborted while Anderson et al. (1982) found that 2 of 26 (7.6%) recipients aborted.

Recent results from our laboratory indicated that certain donor animals did contribute with a high incidence of abortions/resorptions as seen in Table 2. Unfortunately, the aborted material was not suitable for chromosome analysis. It is known, however, that chromosome abnormalities do occur in the embryos (Guyerie, 1983).

Several factors are known to influence the survival rate of ipsilaterally transferred embryos. Besides culture time prior to transfer (Peters et al., 1978; Renard et al., 1980; Schneider et al., 1980), the ease of transfer (Boland, 1976), the skill of the operator (Bowen et al., 1978; Schneider et al., 1980), the instrumentation (Wilmut et al., 1978; Tervit et al., 1978; Jillella & Baker, 1977; Tervit et al., 1980; Takahashi et al., 1981), embryo quality (grade) and developmental stage and donor/recipient and embryo/recipient synchrony (Rowson et al., 1969; Schneider et al., 1980; Wright, 1981; Wright Jr. & Lindner, 1983). A very decisive factor for embryonic survival is the site of transfer in relation to the ovary bearing the CL and contralateral deposition always results in a lower survival rate than ipsilateral transfer (Tervit et al., 1977; Christie et al., 1979; Newcomb & Rowson, 1980; Sreenan et al., 1981; Del Campo et al., 1983).

It has been suggested that lower survival rate of embryos transferred in the horn contralateral to the CL is due to the inability of the embryos to prevent CL regression (Betteridge et al., 1976: Christie et al., 1979; Del Campo et al., 1983).

Unilateral transfers (fig. 1, 2b)

The percentage of pregnancies early in gestation in unilateral tranfers has been found high (about 90%), but the percentage of twins born has been low (approx. 12%). The causes of the loss of pregnancies later in gestation has been attributed to the result of overcrowding in that horn (Rowson et al., 1971).

Bilateral transfers (fig. 1, 2 a)

In case of bilateral transfers a high pregnancy rate also has been reported, however the percentage of twin pregnancies has been quite inconsistent among various studies (see Table 3). This varies between 30-80%.

This is in agreement with naturally occurring multiple pregnancies (Cady and Van Vleck, 1978; Scanlon, 1972; Hanrahan, 1983).

Different laboratories (see table) have been more concerned with improving the technique used in earlier works than studying the causes of embryo loss.

The causes of embryo mortality in bilateral pregnancies has not been critically studied, but is believed to involve reduced uterine capacity or a reduced surface area of contact between the conceptus and the uterine wall (Rowson et al., 1971).

In an experiment designed to study the causes of embryo mortality that occurs in bilaterally pregnant heifers (Del Campo & Ginther, 1982), it was found that direct contact between the embryonic membranes in bilateral pregnancies did not affect the rate of embryo survival before day 50 but the probability that the contact contributes to the loss of embryos later in gestation cannot be excluded.

Williams (1943) suggested that when the fetal sacs are distinct in twin pregnancies one fetus may die but the other may survive. Conversely when the embryonic membranes are fused one fetal death will result in death of the twin.

Ipsilateral embryo versus contralateral embryo - twin transfer

It is known that single embryo transfer in the contralateral horn results in low pregnancy rate (< day 30). However, when another embryo is present in the ipsilateral horn (bilateral pregnancy) the rate of embryonic survival in the contralateral horn increase. According to Del Campo et al.

TABLE 3. Transfer of two embryos.

No. of transfers	Technique	Pregnancy Twin 2x	Single 1x	None 0x	Data collected	% Total rate	Reference	Year
16	S	8	3	5	90 days to calving	69	Rowson et al.	1971
72	S	42	13	17	27-117 days	76	Sreenan et al.	1975
20	S	10	4	6	30 days	75	Sreenan & Beehan	1976
20	N-S	6	6	8	45 days/slaughter	60	Renard et al.	1977
47	N-S	14	13	20	45-120 days	57	Heyman & Renard	1978
40 20	S N-S	7	10	3	42-65 days	85	Newcomb et al.	1978
20	N-S	4	2	14	42-65 days	30	Newcomb et al.	1978
48	S	26	10	12	calving	54	Anderson et al.	1978
77	S	35	14	28	calving	64	Anderson et al.	1979
80	S	41	22	17	42 days/slaughter	70	Newcomb et al.	1980
175	S	62	48	65	calving	57	Anderson et al.	1982
15	S	4	0	11	110 days	27	Del Campo et al.	1983

N-S = Non-surgical transfer
S = Surgical transfer

(1983) it seems that an embryo inserted in the ipsilateral horn would have a protective effect on the contralateral embryo early in gestation. This protective effect would involve prevention of CL-regression by the luteo-tropic effect of the ipsilateral embryo or gravid horn. However, a gradual loss of embryos occurred after day 30 in bilateral transfers.

It appears that the positive effect of the ipsilateral embryo on the survival rate of the contralateral embryo which is present early in gestation (before day 30) becomes negative later in gestation when both embryos are present.

Other authors (Sreenan et al., 1975) have postulated that the higher survival rate of the ipsilateral embryo in bilateral transfers is due to "the effect of the CL on site of embryo transfer". The results of this experiment, however, are difficult to interpret since the "local effect of the ipsilateral embryo on the CL was present" (Del Campo et al., 1977).

It seems to be difficult to separate the "effect of embryos on CL" which has been demonstrated in cattle from "the effect of the CL on side of embryo transfer".

By analysing some of the existing data in the literature the single embryonic survival was found to be 64.4% (380 out of 590 bilaterally transferred embryos survived).

3. TRANSFER OF A SINGLE EMBRYO TO ALREADY MATED RECIPIENTS

Transfer of a single fertilized egg to already mated recipients has been utilized as a means of improving the calf crop. It has been success-fully accomplished in several studies where the day 6-7 egg is deposited in the horn contralateral to the ovary bearing the CL. Either surgical or non-surgical transfer has been performed. The non-surgical transfer is still the predominant method as seen in Table 4.

The twinning percentage has varied from 25-60 (Boland et al., 1975, 1976; Gordon & Boland, 1979; Sreenan & McDonagh, 1979) and the contra-lateral embryo survival is considerably higher (44%, 11/25; Sreenan & McDonagh, 1979) than the contralateral, single transfer to unmated recipients (7%, 4/56; Sreenan et al., 1981).

The survival of either the native or the transferred embryo was 60.7% (258 embryos survived out of 425 inseminated or transferred).

TABLE 4. Transfer of embryos to previously inseminated recipients.

No. of embryos transferred	Technique	Pregnancy			Data collected	Reference	Year
		Twin 2x	Single 1x	None 0x			
24	N-S	6	9	9	30 days	Boland et al.	1975
17	N-S	6	8	3	calving	Testart et al.	1975
6	N-S	2	3	1	90? days to calving	Boland et al.	1976
15	N-S	5	5	5	32/95 days to calving	Boland & Gordon	1978
48	N-S	16	16	16	?	Gordon & Boland	1979
25	N-S	9	6	10	30-42 days	Sreenan & McDonagh	1979
64	S N-S	29	35	0	calving	Holy et al.	1981
84	N-S	20	29	35	calving	Sreenan et al.	1981
112	S	38	20	54	182-196 days	Summer et al.	1983

N-S = Non-surgical transfer

S = Surgical transfer

190

CONCLUSION

Most of the existing literature on non-surgical and surgical transfer
of embryos was analyzed.

Single transfer, twin transfer and transfer to already mated heifers
gave almost identical embryonic survival rates (58.8%, 64.4% and 60.7%,
respectively). Thus, the probability for embryonic survival(P embryo
survival) for whatever method used would be: 5276/8897 = 0.59.

SWINE

Gilts and sows lose naturally a high proportion (40%) of their
embryos (Pope et al., 1982) probably due to overcrowding of the horns
(Dziuk, 1968). Although embryo transfer has not been as widely used in
this species, several studies have been conducted with the study of
embryonic survival following transfer to recipients (Webel et al., 1970).

Recently Pope et al. (1982) transferred embryos to recipients and
found an embryonic survival rate per recipient of 42 $\pm$ 10% and 43 $\pm$ 12%
for 5 and 7 days old embryos, respectively. Similarly, James et al. (1983)
transferred 4-8 cell eggs and morulae and registered an embryonic survival
(%) of 49.5 $\pm$ 13.6 morulae giving a higher survival rate (45.7 $\pm$ 12.2%).

More extensive studies of factors affecting the pregnancy rate and
embryonic survival was undertaken in Cambridge by Polge (1983). He found
that survival following transfer of hatched embryos (day 8/9) was slightly
lower than following transfer of embryos 3-7 days old.

Another interesting point in these studies was the synchronization
experiments. Synchronous transfer resulted in a pregnancy and survival rate
of 75 and 71%, respectively. No decrease in pregnancy or survival rate was
registered in recipients coming in heat 1, 2, or 3 days after the donor.
However, gilts exhibiting heat prior to the donor animals had a considerable reduction in both pregnancy and embryonic survival rate. This is
different from most other species where exact synchrony is required. Polge
also found that the viability of both young and old embryos dropped when
they were cultured for 24 and 48 h.

CONCLUSION

The ever increasing interest for embryo transfer in this species will
probably generate more transfer data and thus sustain the ongoing extensive
in vitro experiments on embryo endocrinology (Niemann & Elsaesser, 1984).

REFERENCES

Alexander, A.M., Marcus, A.N. and Hooton, J.K. 1976. Non-surgical bovine embryo recovery. Vet. Rec., 99, 221.

Anderson, G.B., Cupps, P.T., Drost, M., Horton, M.B. and Wright, R.W., Jr. 1978. Induction of twinning in beef heifers by bilateral embryo transfer. J. Anim. Sci., 46, 449-452.

Anderson, G.B., Cupps, P.T. and Drost, M. 1979. Induction of twins in cattle with bilateral and unilateral embryo transfer. J. Anim. Sci., 49, 1037-1041.

Anderson, G.B., BonDurant, R.H. and Cupps, P.T. 1982. Induction of twins in different breeds of cattle. J. Anim. Sci., 54, 485-490.

Avery, T.L., Fahning, M.L., Pursel, V.G. and Graham, E.F. 1962. Investigations associated with the transplantation of bovine ova. IV. Transplantation of ova. J. Reprod. Fertil., 3, 229-238.

Betteridge, K.J., Mitchell, D., Eaglesome, M.D. and Randall, G.C.B. 1976. Embryo transfer in cattle 10-17 days after estrus. VIIIth Int. Congr. on Anim. Reprod. and A.I., III, 237-240.

Boland, M.P., Crosby, T.F. and Gordon, I. 1975. Twin pregnancy in cattle established by non-surgical egg transfer. Brit. vet. J., 131, 738-740.

Boland, M.P., Crosby, T.F. and Gordon, I. 1976. Birth of twin calves following a simple transcervical non-surgical egg transfer technique. Vet. Rec., 99, 274-275.

Boland, M.P. 1976. Twinning in beef cattle. Surgical and non-surgical approaches. 27th Ann. Meet. Europ. Ass. Anim. Prod., Zürich.

Boland, M.P. and Gordon, I. 1978. Twinning in lactating Friesian cows by non-surgical egg transfer. Vet. Rec., 103, 241.

Bowen, J.M., Elsden, R.P. and Seidel, G.E., Jr. 1978. Non-surgical embryo transfer in the cow. Theriogenology, 10, 89-95.

Brand, A., Aarts, M.H., Zaayer, D. and Oxender, W.D. 1978. Recovery and transfer of embryos by non-surgical procedures in lactating dairy cattle. In: Control of Reproduction in the Cow. A Seminar in the EEC Programme of Coordination of Research on Beef Production held at Galway, Sept. 1977. Current Topics, Vet. Med., I, 281-291.

Cady, R.A. and Van Vleck, L.D. 1978. Factors affecting twinning and effects of twinning in Holstein dairy cattle. J. Anim. Sci., 46, 950-956.

Christie, W.B., Newcomb, R. and Rowson, L.E.A. 1979. Embryo survival in heifers after transfer of en egg to the uterine horn contralateral to the corpus luteum and the effect of treatment with progesterone and HCG on pregnancy rates. J. Reprod. Fertil., 56, 701-706.

Del Campo, M.R., Rowe, R.F., French, L.R. and Ginther, O.J. 1977. Unilateral relationship of embryos and the corpus luteum in cattle. Biol. Reprod., 16, 580-585.

Del Campo, M.R., Rowe, R.F., Chaichareon, D. and Ginther, O.J. 1983. Effect of the relative locations of embryo and corpus luteum on embryo survival in cattle. Reprod. Nutr. Dévelop., 23, 303-308.

Del Campo, M.R. and Ginther, O.J. 1982. Effect of bilateral pregnancies on development of the conceptus in cattle. Arch. Med. Vet., 12, 95-101.

Dziuk, P.J. 1968. Effect of number of embryos and uterine space on embryo survival in the pig. J. Anim. Sci., 27, 673-676.

Gordon, I. and Boland, M.P. 1979. Cattle twins by egg transfer. Irish vet. J., 33, 79-94.

Greve, T. 1981. Bovine egg transplantation in Denmark. Dissertation.

Greve, T. and Lehn-Jensen, H. 1979 . Embryo transplantation in cattle. Non-surgical transfer of 6½-7½ day old embryos to lactating dairy cows under farm conditions. Acta vet. Scand., 20, 135-144.

Guyerie, F. 1983. Personal Comm.

Hahn, J., Moustafa, L.A., Schneider, U., Hahn, R., Romanowski, W. and Roselius, R. 1978. Survival of cultured and transported bovine embryos following surgical and non-surgical transfers. In: Control of Reproduction in the Cow. A Seminar in the EEC Programme of Coordination of Research on Beef Production held at Galway, Sept. 1977. Current Topics, Vet. Med., I, 356-373.

Hahn, J. 1984. Personal Comm.

Hanrahan, J.P. 1983. The inter-ovarian distribution of twin ovulations and embryo survival in the bovine. Theriogenology, 20, 3-11.

Heyman, Y. and Renard, J.P. 1978. Production de sumeaux par transplantation d'embryons chez les bovins de race a viande. Ann. Med. Vet. 122, 157-163.

Holý, L., Jiřícek, A., Vanatka, F., Vrtel, M. and Fernandez, V. 1981. Artificial induction of twinning in cattle by means of supplemental embryo transfer. Theriogenology, 16, 483-487.

James, J.E., James, D.M., Martin, P.A., Reed, D.E. and Davis, D.L. 1983. Embryo transfer for conserving valuable genetic material from swine herds with pseudorabies. J. Amer. vet. med. Ass., 183, 525-528.

Jillella, D. and Baker, A.A. 1977. Transcervical transfer of bovine embryos. Vet. Rec., 103, 574-576.

Lampeter, W.W., Kruff, B., Rieger, J., Wagner, H.G. and Ertel, J. 1980. Beschreibung eines Verfahrens zum unblutigen Transfer boviner Embryonen. Zuchthyg., 15, 173-176.

Lindner, G.M. and Wright, W., Jr. 1983. Bovine embryo morphology and evaluation. Theriogenology, 20, 407-416.

Markette, K.L., Seidel, G.E., Jr. and Elsden, R.P. 1980. Embryonic loss after bovine embryo transfer. Theriogenology, 13, 105.

Newcomb, R., Christie, W.B. and Rowson, L.E.A. 1978. Comparison of the fetal survival rate in heifers after the transfer of an embryo surgically to one uterine horn and non-surgically to the other. J. Reprod. Fertil., 52, 395-397.

Newcomb, R. 1979. Surgical and non-surgical transfer of bovine embryos. Vet. Rec., 105, 432-434.

Newcomb, R. and Rowson, L.E.A. 1980. Investigation of physiological factors affecting non-surgical transfer. Theriogenology, 13, 41-49.

Newcomb, R., Christie, W.B. and Rowson, L.E.A. 1980. Fetal survival rate after the surgical transfer of two bovine embryos. J. Reprod. Fertil, 59, 31-36.

Nielsen, L. and Østergaard-Nielsen, N. 1979. Non-surgical transplantation of seven day old cattle embryos to cows and heifers. Dansk Vet.-Tidsskr., 62, 972-973.

Niemann, H. and Elsaesser, F. 1984. Uptake and effects of ovarian steroids in the early pig embryo: in vitro and in vivo studies. Theriogenology, 21, 84.

Peters, D.F., Anderson, G.B., BonDurant, R., Cupps, P.T. and Drost, M. 1978. Transfer of cultured bovine embryos. Theriogenology, 10, 337-342.

Polge, C. 1983. Embryo transplantation in the pig. In: 2nd World Conf. on Embryo Transfer and in Vitro Fertilization in Mammals. Merieux, Ch. and Bonneau, M. (eds.), 235-242.

Pope, W.F., Maurer, R.R. and Stormshak, F. 1982. Survival of porcine embryos after asynchronous transfer. Proc. Soc. Exp. Biol. Med., 171, 179-183.

Renard, J.P,, Heyman, Y. and du Mesnil du Buisson, F. 1977. Unilateral and bilateral cervical transfer of bovine embryos at the blastocyst stage. Theriogenology, 7, 189-194.

Renard, J.-P., Heyman, Y. & Ozil, J.-P. 1980. Importance of gestation losses after non-surgical transfer of cultured and non-cultured bovine blastocysts. Vet. Rec., 197, 152-153.

Rowe, R.F., Del Campo, M.R., Chritser, J.K. and Ginther, O.J. 1980. Embryo transfer in cattle: Non-surgical transfer. Amer. J. vet. Res., 41, 1024-1028.

Rowson, L.E.A. and Moor, R.M. 1966. Non-surgical transfer of cow eggs. J. Reprod. Fertil., 11, 311-312.

Rowson, L.E.A., Moor, R.M. and Lawson, R.A.S. 1969. Fertility following egg transfer in the cow; effect of method, medium and synchronization of oestrus. J. Reprod. Fertil., 18, 517-523.

Rowson, L.E.A., Lawson, R.A.S. and Moor, R.M. 1971. Production of twins in cattle by egg transfer. J. Reprod. Fertil., 25, 261-269.

Scanlon, P.F. 1972. Frequency of transuterine migration of embryos in ewes and cows. J. Anim. Sci., 34, 791-794.

Schneider, H.J., Jr., Castleberry, R.S. and Griffin, G.L. 1980. Commercial aspects of bovine embryo transfer. Theriogenology, 13, 73-85.

Sreenan, J.M., Beehan, D. and Mulvehill, P. 1975. Egg transfer in the cow: Factors affecting pregnancy and twinning rates following bilateral transfers. J. Reprod. Fertil., 44, 77-85.

Sreenan, J.M. and Beehan, D. 1976. Effects of site of transfer on pregnancy and twinning rates following bilateral egg transfer in the cow. J. Reprod. Fertil., 48, 223-224.

Sreenan, J.M. and McDonagh, T. 1979. Comparison of the embryo survival rate in heifers following artificial insemination, non-surgical blastocyst transfer or both. J. Reprod. Fertil., 56, 281-284.

194

Sreenan, J.M., Diskin, M.G. and McDonagh, T. 1981. Induction of twin-calving by non-surgical embryo transfer: A field trial. Vet. Rec., 109, 77-80.

Sugie, T. 1965. Successful transfer of a fertilized bovine egg by non-surgical techniques. J. Reprod. Fertil., 10, 197-201.

Summers, P.M., Shelton, J.N. and Edwards, J. 1983. The production of mixed-species bos taurus-bos indicus twin calves. Anim. Reprod. Sci., 6, 79-89.

Takahashi, Y. 1980. Comparative studies on bovine embryo transfer for three different techniques of transcervical, flank surgical and cervical bypass methods. Japan. J. Anim. Reprod., 26, 155-157.

Takahashi, Y., Suzuki, T. and Saito, N. 1981. Cervical bovine embryo transfer with ensheathed metal AI instrument. Japan. J. Anim. Reprod., 27, 54-55.

Takahashi, Y., Suzuki, T. and Shimohira, I. 1982. Experiments on cervical embryo transfer in cattle. Effects of paper sheath covering transfer instruments on pregnancy rates and uterine bacterial contaminations. Japan. J. Anim. Reprod., 28, 30-35.

Tervit, H.R., Havik, P.G. and Smith, J.F. 1977. Egg transfer in cattle: Pregnancy rate following transfer to the uterine horn ipsilateral or contralateral to the functional corpus luteum. Theriogenology, 7, 3-8.

Tervit, H.R., Goold, P.G. and Cooper, M.W. 1978. Non-surgical embryo transfer in cattle. N.Z. vet. J., 26, 215.

Tervit, H.R., Cooper, M.W., Goold, P.G. and Haszard, G.M. 1980. Non-surgical embryo transfer in cattle. Theriogenology, 13, 63-71.

Testart, J., Godar-Siour, C. and du Mesnil du Buisson, F. 1975. Trans-vaginal transplantation of an extra egg to obtain twinning in cattle. Theriogenology, 4, 163-168.

Trounson, A.O., Rowson, L.E.A. and Willadsen, S.M. 1978 . Non-surgical transfer of bovine embryos. Vet. Rec., 102, 74-75.

Webel, S.K., Peters, J.B. and Anderson, L.L. 1970. Synchronous and asynchronous transfer of embryos in the pig. J. Anim. Sci., 30, 565-568.

Williams, W.L. 1943. Diseases of the genital organs of domestic animals. 3rd Ediţ., Plimpton, E.W. (ed.), Ithaca, N.Y.

Wilmut, I., Sales, D.I., Manson, C. and Newell, G. 1978. Non-surgical transfer of cattle embryos - A field trial. ARC Report, Edinburgh, 41.

Wright, J.M. 1981. Non-surgical embryo transfer in cattle. Embryo-recipient interactions. Theriogenology, 15, 43-56.

THE ROLE OF THE MALE IN EMBRYONIC MORTALITY
(Cattle and Sheep)

M. Courot, G. Colas

I.N.R.A., Station de Physiologie de la Reproduction
37380 NOUZILLY (F)

ABSTRACT
      The male contributes not only to the fertilization of females but also, through the quality of spermatozoa, to the capacity of embryo to survive to become viable offspring. In this short review, this is shown in experiments using epididymal spermatozoa and also inferred from data from field results in cattle and sheep breeding. Factors involved in these processes are the genetic make up (breeds, strains, chromosomal abnormalities), environment (photoperiod, temperature) and management (semen quality, semen handling) of the males. Most of the male contribution to embryonic loss operates soon after fertilization and tools are needed to detect pregnancy very early to allow diagnosis of the absence of fertilization and therefore very early embryonic mortality.

INTRODUCTION

      The fertility of domestic mammals can be estimated at various intervals after fertilization. It is known that, in all species, the results vary with the interval between coitus or artificial insemination (AI) and the time of measurement of fertility. Although the chances of fertilization (penetration of an egg by a spermatozoon and subsequent cleavage of the egg) are generally high, being more than 90% in normal conditions (Kidder et al., 1954; Bearden et al., 1956; Killeen, 1974; Kilgour and Wilkins, 1980), it is common to see a given percentage of females again showing oestrus one oestrus cycle after mating or A.I. due to failure of fertilization or to early embryonic mortality. Later on, there is a further drop in the proportion of females pregnant due to embryonic and foetal mortality.

      What is the part played by the male in this problem? The contribution of the male to fertility in mammals is not restricted to the simple fertilization of eggs. The quality of fertilization must be such as to promote the normal development of the conceptus and survival of the embryo. If not, the embryo will degenerate, resulting in the embryonic mortality which occurs more or less rapidly after fertilization. It is known that not only the female but also the male contributes to fertilization and embryonic mortality.

We have been asked to review the role of the male in embryonic mortality in cattle and sheep. To do this, we shall first present data showing the male contribution to the problem. These data are taken from experiments and from studies of reproduction in the field. Finally, we shall consider some factors involved in this male deficiency.

THE MALE CONTRIBUTION TO EMBRYONIC MORTALITY : ITS EXISTENCE
Experimental data

These are given to illustrate, in purely experimental conditions, far from practical application, how the quality of spermatozoa could interfere with embryonic survival.

In studies on epididymal sperm maturation in the ram, spermatozoa were collected from different parts of the epididymis (head, body, tail) and used, in comparison with ejaculated spermatozoa, to fertilize ewes by intrauterine surgical insemination. (This technique was used in order to avoid the problem of the inability of non-motile spermatozoa to pass through the cervix). Results were analyzed for different intervals after insemination, namely fertilization and cleavage of eggs 38 or 48h post A.I., diagnosis of pregnancy by progesterone assay in the peripheral blood at 18 d. post A.I. and lambing.

Immature spermatozoa taken from body of the epididymis were able to fertilize some ovocytes, but a lower proportion than mature spermatozoa (30 vs 85 %), and to initiate embryonic development. However, the kinetics of the first cleavages of eggs was abnormal, fertilization with immature spermatozoa resulting in delayed segmentation (Fournier-Delpech et al., 1981; Tables 1 and 2). This did not result in normal pregnancy, since no ewes inseminated with these immature spermatozoa lambed (Fournier-Delpech et al., 1979; Table 3). With mature spermatozoa from the tail of the epididymis or from the ejaculate, most if not all females with a positive pregnancy diagnosis at 18 days gave birth to normal lambs. These experimental results clearly show a specific effect of the male spermatozoon on embryonic mortality in the sheep.

Similar results have previously been obtained in rabbits (Orgebin-Crist and Jahad, 1977). All these data taken together show that this effect is not restricted to species with high prolificacy, but that it can be considered as general with some contributions of male origin occurring very early and others later in embryonic development.

197

TABLE 1   Sperm maturation and fertilization of eggs in ewes inseminated
           with spermatozoa of different origin

Origin of spermatozoa	Ewes No	Eggs[1] collected No %[2]	cleaved No (%)
Epididymal Tail or Ejaculate	24	34 (64)	29 (85)
			P<0.01
Epididymal Body	16	27 (64)	8 (30)

(1) observed 38 and 48h post A.I.   (Fournier-Delpech et al., 1981)
(2) No Eggs/No ovulations

TABLE 2   Sperm maturation and cleavage of eggs in ewes inseminated with
           spermatozoa of different origin

Origin of spermatozoa	Cleaved eggs: number cells 38h post A.I.		48h post A.I.		
	2cells	4cells	2cells	4cells	>4cells
Epididymal Tail or Ejaculate	17/21 80%	4/21 19%	1/8 13%	0/8 0%	7/8 87%
Epididymal Body	4/4 100%	0/4 0%	1/4 25%	3/4 75%	0/4 0%

(Fournier-Delpech et al., 1981)

TABLE 3   Sperm maturation and embryonic loss in ewes inseminated with
           spermatozoa of different origin

Origin of spermatozoa	Pregnancy[1] 18d. post A.I. No %		Embryonic[2] survival No %		Embryonic loss %
distal head	0/8	0	-	-	-
median body	6/39	15	0/6	0	100
distal body	19/35	54	15/19	77	23
proximal tail	29/37	78	25/29	86	14
distal tail	16/20	80	16/16	100	0

(Fournier-Delpech et al., 1979)
(1) Progesterone Assay in the blood plasma (Terqui and Thimonier, 1974)
(2) No females lambing/No females pregnant at 18d.

Data from field results in A.I.

Bovine - A series of papers published between 1950 and 1960 on bovine fertility after A.I. have already shown the existence of significant differences in fertilization rate of bulls known as having high fertility (100%) or low fertility (71%) (Kidder et al., 1954). Moreover, when analysing the return rate of females inseminated with semen from such bulls, there appeared to be a higher embryonic mortality in bulls of low fertility than in those of high fertility (19.2 vs 10.5% for embryonic mortality measured at 33 days post A.I. (Bearden et al., 1956). More generally, it has been shown that the higher the fertilization rate, the lower the embryonic wastage (Courot and Tourneur, 1976). This is shown by a high non-return (NR) rate at 29 days post A.I. and very little difference between this NR and the 180d NR for the bulls of higher fertility compared to the reverse situation for the bulls with lower fertility (Erb and Flerchinger, 1954) (Table 4).

TABLE 4   Embryonic mortality following insemination with semen from bulls of differing fertilizing ability

Groups of bulls with increasing fertility	No 1st services	Pregnancy		Embryonic loss	
		29d(A)	180(B)	A-B	% $\frac{(A-B)}{A}$
A	6210	70.4	53.3	17.1	24.3
B	10030	74.0	58.6	15.2	20.8
C	18879	76.7	63.1	13.6	17.7
D	17164	79.8	67.0	12.8	16.0

(Erb and Fleschinger, 1954)

The negative correlation between the level of early NR rate and the difference between early and late NR rate has been largely shown using fresh and frozen bull semen. However, comparing pregnancy diagnosis by NR rate or by blood progesterone levels, Humblot et al. (1982, 1983) consider that, in dairy cattle, there is no further influence of the male on embryonic loss after 24 days post A.I.. They also believe that embryonic mortality could have been overestimated due to so called "post-insemination anoestrus", that is females with delayed return to oestrus in spite of a low level of blood progesterone 21 days post A.I. (case B in Fig. 1).

This problem requires more experimentation and analysis in relation to the variability of the level of progesterone during early pregnancy, around the period of "normal" return to oestrus, and occurrence of oestrus (Thirapatsukun et al., 1978; Nakao et al., 1983; Fig. 1).

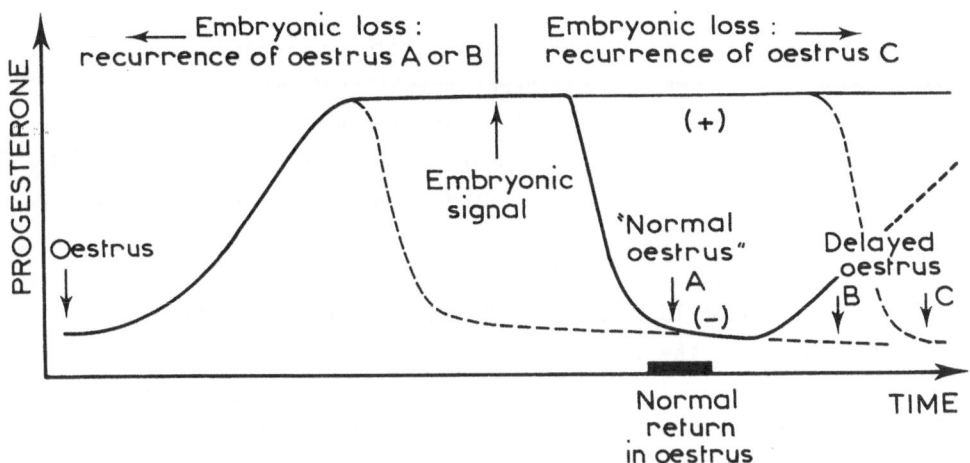

Fig. 1 Progesterone level in the peripheral blood or in milk of the cow or ewe in relation to the normal oestrus cycle or pregnancy. After mating or A.I. at the first oestrus, the female is either fertilized or not. * If not, at the time of normal luteolysis progesterone decreases and a new "normal" oestrus occurs (case A). The normal range of variability of its occurrence is shown on the abcissa (ie 18-24 days in cattle).
* If fertilized, the progesterone pattern depends on the subsequent development of the embryo
- normal: progesterone remains elevated throughout pregnancy
- embryonic loss before embryonic signal (ie 16 days in cattle): progesterone decreases as in absence of fertilization, oestrus occurs normally as in case A.
- embryonic loss after embryonic signal: delayed decrease in progesterone, oestrus occurs as in case C.

Problems
1) with progesterone assay, it is not possible to differentiate between absence of fertilization and early embryonic loss.
2) delayed oestrus (case B) is sometimes observed in spite of a low level of progesterone during the period of "normal" return to oestrus, (- on the graph). It is related to post A.I. anoestrus (Humblot et al., 1983) or to the variability of progesterone levels compatible with pregnancy (Nakao et al., 1983); in this case, this is a "false" negative diagnosis of pregnancy. Without a reliable very early pregnancy diagnosis, one cannot decide what is the situation.

Sheep - Embryonic mortality also exists in sheep, the mean rate falling between 20 and 30% (Edey, 1969) with individual ram variation (Blockey et al., 1975). The effect of the male was suggested in experiments using rams selected for high and low prolificacy showing estimated embryonic survival rates of 81 and 89% respectively in ewes mated to low and high line rams (Burfening et al., 1977). It was also shown by the positive correlation observed between prolificacy and lambing rate obtained with semen of different rams (Despierres et al., 1981). A higher embryonic loss was also observed in some experiments with deep frozen semen (Langford et al., 1979), especially when semen was deposited directly into the uterus (Salamon and Lightfoot, 1967; Lightfoot and Salamon, 1970).

THE MALE CONTRIBUTION TO EMBRYONIC MORTALITY : FACTORS INVOLVED

When looking at the literature dealing with embryonic death in cattle and sheep, one is surprised to note that most work has been devoted to the role of the female, rarely to that of the male. However it is clear from the above that the male can play an important role in the survival of the foetus in the uterus. Moreover the relative weight of individual males on the fertility of herds is considerable compared to that of individual females. It is thus interesting to review the main causes of embryonic mortality due to the male genitor. This analysis will not deal with the pathological factors affecting semen quality of the male, which would be worthy of a specific study.

Genetic factors

It is generally known that such chromosomal abnormalities as translocation, deletion and trisomy contribute to early embryonic mortality. This has been shown especially in the bull (Gustavsson, 1975; 1979) in the pig (Ackerson and Henricson, 1972) and in man (Mattei, 1984), and while the same phenomenon is likely to exist in the ram, no precise study has been yet done in this species.

Genetic origin of strain seems to modify embryonic mortality in the sheep. Burfening et al. (1977) obtained levels of embryonic mortality of 11 and 19 % respectively from ewes mated to rams born from dams selected for high and low prolificacy. Conversely, Barker and Land (1970) found no differences in prolificacy between rams of two breeds which differed markedly in litter size, so that the ability of their rams to modify embryo survival was probably the same. Therefore, the possibility of an effect of genetic origin of the ram has not been conclusively resolved.

Environmental factors

Photoperiodism

Several experiments have shown the importance of photoperiod in the control of the sexual activity in the ram (Ortavant, 1958; Pelletier, 1971; Lincoln and Davidson, 1977; Pelletier et al., 1981). In this species there are also large seasonal variations in the fertilizing capacity and embryonic death rates.

Furthermore, when ewes were inseminated at the same time with semen from rams exposed to either short or long days, embryonic mortality, estimated as the difference between pregnancy rate at 18 days post A.I. and lambing rate, was higher in ewes inseminated with semen from the rams exposed to long days (Colas, 1983; Table 5). These results suggest, as previously indicated for the bull, that the lower the lambing rate, the higher the embryonic loss.

TABLE 5   The effect of photoperiodic treatment of rams on embryo mortality (1)

Ram	Pregnancy rate 18 days post AI (%)(2)	Lambing rate (%)	Embryonic loss (%)
Long day rams (increasing light)	63.5(b) —P<0.05— (104)	47.0 (104)	26.0
Short day rams (decreasing light)	66.0 ——NS—— (119)	57.1 (119)	13.0

(Colas, 1983)

(1): ewes from the same flocks inseminated at the same time, 55 hours following FGA + PMSG treatment.
(2): from unselected ejaculates.

Moreover, semen quality also is worse in animals exposed to long days (Colas, 1983); this shows that embryo mortality directly depends on the initial quality of the gametes. If so, it is likely that embryonic loss must be higher in the very young ram lamb, whose sperm contains more morphologically abnormal spermatozoa than in the adult (Skinner and Rowson, 1968; Colas and Zinszner-Pfimlin, 1975). A similar effect of age has been shown on the reproductive performance in females, younger ewes performing worse than older ewes (Restall et al., 1975; Quirke and Hanrahan; 1977).

Ambient temperature

The detrimental effect of high ambient temperature on the quality of male gametes is well established. Exposure of rams to hot temperatures for a long enough time results in an increase of embryonic mortality (Rathore, 1968; Table 6).

TABLE 6   Heat treatment of rams and fertility of ewes to which they were mated

Ewe performance	Duration of heat treatment of rams		
	Control	2 days	4 days
23 day non-return rate	14/20	9/20	3/20
40 day pregnant*	12/20	3/20	0/20
ewes lambed*	12/20	3/20	0/20

* all groups significantly different.   (Rathore, 1968)

As can be seen, the difference in fertility between control rams and rams subjected to two days of treatment is significant only from the 40th day of pregnancy. In cattle, the effect of hot climate seems to affect the cow rather than the bull (Stott and Williams, 1962).

Semen quality

The quality of semen has been shown to influence embryonic survival, especially in the bull. These studies do not show whether such embryonic mortality is due only to the poor quality of ejaculates or to a genetic male factor. However, it has been clearly shown in cattle that embryonic loss can be suppressed to a great extent with appropriate semen technology, such as the addition of antibiotics to the milk diluent (see review by Salisbury et al., 1978).

Technological handling of sperm

Storage conditions of diluted semen often have a deleterious effect on embryo survival. This has been clearly established in the ram by Maxwell (1978) who demonstrated that the percentage of embryonic mortality increased with the period of storage at 4°C (Table 7).

The time of mating or insemination according to heat or ovulation, is also important for success of A.I. through fertilization rate and embryonic loss (see review by Salisbury et al., 1978) but this is more related to management than a specific male effect on reproduction.

TABLE 7   Effect of storage of ram semen on its fertility

Age of semen (days)	Pregnancy rate % ( )		lambing rate %	embryonic loss %
0 (fresh)	70.0	(70)	60.0	14.3
1	45.7	(70)	34.3	25.0
2	46.5	(71)	33.8	27.3
3	32.9	(70)	17.1	47.8

( ) ewes inseminated.   (Maxwell, 1978)

CONCLUSION

It is evident from these data that the male contributes not only to the fertilization of females but, through the quality of spermatozoa, to the capacity of embryo to survive and to give normal viable offspring. This could explain the correlation between fertilization rate and prolificacy and the direct effect of the male on the litter size in sheep and on fertility in cattle. Most embryonic mortality occurs early in pregnancy (see other chapters in this book) but, at the present time, one is still unable to distinguish between absence of fertilization and early embryonic loss when the latter occurs before the signal has been given by the pregnant uterus for the maintenance of a corpus lutueum. This points out the imperative need for such an accurate diagnosis of early pregnancy as EPF (Morton et al., 1979; Georgieva and Stefanov., 1984 ; Koch et al., this meeting) or any other efficient technique.

Several factors affect embryonic loss due to the male's contribution. Most of these operate through the impaired quality of the male gametes which is really a key-point in this problem. All efforts made to improve quality of spermatozoa such as better animal care or environment, well adapted semen technology or selection will, at the same time, improve fertilizing capacity and embryonic survival rate, which is of high economic importance in farm animal industries.

204

ACKNOLEDGMENTS

The authors are indebted to Dr. R. ORTAVANT and Mr. R.J. KILGOUR for critical discussion of the manuscript and to Mr. R.J. KILGOUR for improvment of the English.

REFERENCES
Ackerson, A., Henricson, B. 1972. Embryonic death in pigs caused by unbalanced karyotype. Acta Vet. Scand., 13, 151-160.
Barker, J.D., Land, R.B. 1970. A note on the fertility of milk ewes mated to Finnish Landrace and Border Leicester rams. Anim. Prod., 12, 673-675.
Bearden, H.J., Hansel, W.M., Bratton, R.W. 1956. Fertilization and embryonic mortality rates of bulls with histories of either low or high fertility in artificial breeding. J. Dairy Sci., 39, 312-318.
Burfening, P.J., Friedrick, R.L., Van Horn, J.L. 1977. Estimates of early embryonic loss in ewes mated to rams selected for high and low prolificacy. Theriogenology., 7, 285-291.
Colas, G. 1983. Factors affecting the quality of ram semen. In "Sheep Production" (Ed. W. Haresign). (Butterworths). pp. 453-465.
Colas, G., Zinszner-Pfimlin, F. 1975. Production spermatique et développement testiculaire chez l'agneau de race Ile-de-France et Préalpes. In 1ères Journées de la Recherche Ovine et Caprine; "les races prolifiques" (Ed. Itovic). pp. 235-243.
Courot, M., Tourneur, J.C. 1976. Qualité du sperme et réussite de l'insémination artificielle. In "Maîtrise des cycles sexuels chez les bovins" (Ed. Inra/Sersia/Searle, Paris). pp. 117-123.
Despierres, J., Lemaire, M., Lambert, J., Colas, G. 1981. cité par Colas, G., Menissier, F., Courot, M., Paquignon, M. 1984. In "Insémination artificielle et amélioration génétique; bilan et perpectives critiques". Toulouse-Auzeville (France), 23-24 Novembre 1983 (Ed. Inra Publications, colloque Inra 29) (in press).
Edey, T.N. 1969. Prenatal mortality in sheep: a review. Anim. Breed. Abstr. 37, 173-190.
Erb, R.E., Flerchinger, F.H. 1954. Influence of fertility level and treatment of semen on non-return decline from 29 to 180 days following artificial service. J. Dairy Sci., 37, 938-948.
Fournier-Delpech, S., Colas, G., Courot, M., Ortavant, R., Brice, G. 1979. Epididymal sperm maturation in the ram: motility, fertilizing ability and embryonic survival after uterine artificial insemination in the ewe. Annls. Biol. anim. Biochim. Biophys., 19, 597-605.
Fournier-Delpech, S., Colas, G., Courot, M. 1981. Observations sur les premiers clivages des oeufs intratubaires de brebis après fécondation avec des spermatozoïdes épididymaires ou éjaculés. C.R. Acad. Sci. Paris, série D, 292, 515-517.
Georgieva, R., Stefanov, D. 1984. Early pregnancy factor (EPF) in cows and sows. Detection of pregnancy, embryonic mortality and its partial characterization. 10th Int. Congr. Anim. Reprod and A.I., 2, Abst 83.
Gustavsson, I. 1975. New information on the reduced fertility of cattle with the 1-29 translocation. Eur. coll. über zytogenetik in Veterinärmedizin, Tierzucht und Saügetierkund. Giessen p. 184.
Gustavsson, I. 1979. Distribution and effects of the 1-29 Robertsonian translocation in cattle. J. Dairy Sci., 62, 825-835.
Humblot, P., Della Porta, M.A., Schwartz, J.L. 1982 and 1983. Etude de la mortalité embryonnaire. Elev. and Insem., 189, 15-28 and 194, 3-1?.

Kidder, H.E., Black, W.G., Wiltbank, J.N., Ulberg, L.C., Casida, L.E. 1954. Fertilization rates and embryonic death rates in cows bred to bulls of different levels of fertility. J. Dairy Sci., 37, 691-697.

Kilgour, R.J., Wilkins, J.F. 1980. The effect of serving capacity of the ram syndicate on flock fertility. Austr. J. Exp. Agric. Anim. Husb., 20, 662-666.

Killeen, I.D. 1974. Effects of liveweight of Border-Leicester x Merino ewes and breed of ram on fertilization in flock-mated sheep. J. Reprod. Fert., 36, 464-465.

Langford, G.A., Marcus, G.J., Hackett, A.J., Ainsworth, L., Wolynetz, M.S., Peters, H.F. 1979. A comparison of fresh and frozen semen in the insemination of confined sheep. Can. J. Anim. Sci., 59, 685-691.

Lightfoot, R.J., Salamon, S. 1970. Fertility of ram spermatozoa frozen by the pellet method. II. The effects of method of insemination on fertilization and embryonic mortality. J. Reprod. Fert. 22, 399-408.

Lincoln, G.A., Davidson, W. 1977. The relationship between sexual and aggressive behaviour and pituitary and testicular activity during the sexual cycle of rams, and the influence of photoperiod. J. Reprod. Fert., 49, 267-276.

Mattei, J.F. 1984. Influence of the male in embryonic mortality. In "The male in farm animal reproduction" (Ed. M. Courot, M. Nijhoff publ. for the CEC) pp. 350-369.

Maxwell, W.M.C. 1978. Studies on the survival and fertility of chilled-stored ram spermatozoa and frozen stored boar spermatozoa. PhD Thesis. University of Sydney.

Morton, H., Nancarrow, C.D., Scaramuzzi, R.J., Evison, B.M., Clunie, G.J.A. 1979. Detection of early pregnancy in sheep by the rosette inhibition test. J. Reprod. Fert, 56, 75-80.

Nakao, T., Sugihashi, A., Kawata, K., Saga, N., Tsunoda, N. 1983. Milk progesterone levels in cows with normal or prolonged estrous cycles, referenced to an early diagnosis. Jap. J. Vet. Sci., 45, 495-499.

Orgebin-Crist, M.C., Jahad, N. 1977. Delayed cleavage of rabbit ova after fertilization by young epididymal spermatozoa. Biol. Reprod., 16, 358-362.

Ortavant, R. 1958. Le cycle spermatogénétique chez le bélier. Thèse Doct. Sci. Paris 127 pp. ed 1959. Ann. Zootech., 8, 183-244 and 271-322.

Ortavant, R. 1977. Photoperiodic regulation of reproduction in the sheep. Proc. "Management of reproduction in Sheep and Goats symposium (Univ. of Wisconsin, Madison)" pp. 58-71.

Pelletier, J. 1971. Influence du photopériodisme et des androgènes sur la synthèse et la libération de LH chez le bélier. Th. Doct. Etat ès Sciences nat. Paris Cnrs, A.D. 5441.

Pelletier, J., Blanc, M., Daveau, A., Garnier, D.H., Ortavant, R., de Reviers, M.M., Terqui, M. 1981. Mechanism of light action in the ram: a photosensitive phase for LH, FSH, testosterone and testis weight. In "Photoperiodism and Reproduction in vertebrates" (Nouzilly, France 24-25 Septembre 1981). Int. Coll. INRA 6, 117-134.

Quirke, J.F., Hanrahan, J.P. 1977. Comparison of the survival in the uteri of adult ewes of cleaved ova from adult ewes and ewe lambs. J. Reprod. Fert., 51, 487-489.

Rathore, A.K. 1968. Effects of high temperature on sperm morphology and subsequent fertility in Merino sheep. Proc. Aust. Soc. Anim. Prod., 7, 270-274.

Restall, B.J., Wilkins, J., Kilgour, R.J., Tyrrell, R.N., Hearnshaw, H. 1976. Assessment of reproductive wastage in sheep. 3. An investigation of a commercial sheep flock. Austr. J. Exp. Agric. Anim. Husb., 16, 344-352.

Salamon, S., Lightfoot, R.J. 1967. Fertilization and embryonic loss in sheep after insemination with deep frozen semen. Nature, 216 (5111), 194-195.

Salisbury, G.W., Van Demark, N.L., Lodge, J.R. 1978. Physiology of reproduction and artificial insemination of cattle. 2nd ed. (Freeman and Co, San Francisco).

Skinner, J.D., Rowson, L.E.A. 1968. Puberty in Suffolk and cross-bred rams. J. Reprod. Fert., 16, 479-488.

Stott, G.H., Williams, R.J. 1962. Causes of low breeding efficiency in dairy cattle associated with seasonal high temperatures. J. Dairy Sci., 45, 1369-1375.

Terqui, M., Thimonier, J. 1974. Nouvelle méthode radioimmunologique rapide pour l'estimation du niveau de progestérone plasmatique : application chez la brebis et chez la chèvre. C.R. Acad. Sci. Paris, 279, 1109-1112.

Thirapatsukun, T., Entwistle, K.W., Gartner, R.J.W. 1978. Plasma progesterone levels as an early pregnancy test in beef cattle. Theriogenology, 9, 323-328.

GENETIC CONSTITUTION OF EARLY STAGE PIG EMBRYOS AND EMBRYO MORTALITY

P. de Boer, F.A. van der Hoeven and M.P. Cuijpers

Dept. of Genetics, Agricultural University,
Gen. Foulkesweg 53, 6703 BM, Wageningen,
The Netherlands

ABSTRACT

The results of cytogenetic studies in pre- and postimplantation pig embryos have been compared with results from sheep and cattle, experimental rodents such as mouse, chinese hamster and rabbit, and with man. It can be concluded that of the categories of chromosome mutants that are mostly of concern for embryonic death, aneuploidy and polyploidy, man has the highest incidence of both. For the other species, levels of both types of abnormalities generally are low with the exception of aneuploidy for sheep embryos and possibly polyploidy for certain aspects of pig reproduction (delayed fertilization). Assessments of the contribution of chromosome abnormalities to embryonic death are 30% for the mouse, which generally has a low level of embryonic death, $\sim$50% in man with a high level and "low" for the pig with likewise a high level of embryonic death.

INTRODUCTION

Research into the genetic contribution to embryonic and foetal death has mainly been carried out in laboratory rodents such as the chinese hamster, the rabbit and the mouse. Epidemiological research in human abortus material is another source of knowledge concerning the genetic aspects of embryonic and foetal death. In both experimental rodents and in man, the emphasis is entirely on chromosome mutations, of which two classes are important in this context.

The first class of aneuploid embryos arises through non-disjunctional events during one of the two meiotic divisions in either oogenesis or spermatogenesis. At the first meiotic division, non-disjunction denotes the failure of homologous chromosomes to orientate properly such that each division pole receives one copy of the chromosome(s) concerned. At the second meiotic division, non-disjunction means an incorrect distribution of the two chromatids of one or more chromosomes over the two poles. Both processes lead to gametes with incorrect chromosome numbers in secondary oocyte and/or female pronucleus and in the fertilizing spermatozoon, the genetic effect being expressed as a dominant lethal during prenatal development.

The second class comprises of genome mutations: the presence of extra genomes (haploid sets of chromosomes) or the lack of a genome which results in an embryo with a haploid chromosome complement.

When one extra genome is present, it can be derived through an error in female meiosis (first or second meiotic division, digyny), in male meiosis (first or second meiotic division, diandry) or through fertilization by two spermatozoa (dispermy). Haploid and diploid parthenogenotes, that both are prenatal lethals, can also be grouped in this category.

Besides dominant lethals, prenatal death can theoretically also be caused by homozygosity for a recessive lethal genetic factor. For purposes of radiation risk estimates, using the mouse as the experimental model, the mutation rate per genome for recessive lethal factors has been estimated to be 0.5% (Lüning, 1975). Thus 0.25% = 1 in 400 conceptuses should die because of homozygosity for such a factor. Because of the low rate, and because of the fact that there is no formal proof for the presence of recessive lethal factors in farm animals, we will leave it out of consideration here.

In the following paragraphs we will focus our attention on the knowledge of chromosome mutations in pig embryos and give reference to the knowledge concerning this topic in other farm animals, experimental rodents and in man.

THE INCIDENCE OF CHROMOSOME ABNORMALITIES IN PIG CONCEPTUSES

Table 1 gives the results reported to date in the literature for different stages of pregnancy. A first impression is that the presence of an extra chromosome (or a chromosome missing) is a rare event in pig preimplantation embryos. Theoretically, the primary incidence of this category of chromosome mutation (and also of polyploidy) should be measured at the first cleavage division, because a priori, nothing is known about the period of embryonic development when chromosome abnormalities do lead to prenatal death in the pig. However, the study of the genetically imbalanced embryos out of matings between a heterozygote for a reciprocal translocation and a karyologically normal individual, for which pig material is available, should give some insight into this problem. Somewhat over 50% of the gametes from reciprocal translocation heterozygotes carry duplications combined with deficiencies, usually for the exchanged chromosome segments that are typical for reciprocal translocations. These gametes do function properly, as with aneuploid gametes, and the consequences of the genetic imbalance usually are expressed as lethality during embryonic de-

velopment. It has been concluded that mouse and human reciprocal translocations lead to embryonic death in roughly comparable periods i.e. from just
before implantation to the growth of the primitive streak post-implanta-
tion (de Boer and de Maar, 1976).

Three reciprocal translocations in the pig are available for study.
From these data, it can be concluded that for the 11p; 15q translocation,
the majority of genetically imbalanced embryos is still alive 10-11 days
after breeding (Akesson and Henricson, 1972). Some imbalanced types per-
sist until the days 23-28 after breeding, i.e. after implantation. For the
4q+; 14q- pig translocation, death probably starts around the 9th day of
embryonic development (Popescu and Boscher, 1982). For a variety of embryonic ages (days 1-16) King et al. (1981) found some evidence for a
shortage of duplication/deficiency type embryos from the 13q-; 14q+ pig
reciprocal translocation. Taking the long preimplantation period of the
pig into account (when compared to mouse and man), the developmental profiles of chromosomally imbalanced embryos seem to be comparable between
the species here mentioned.

Gropp (1981) has most accurately studied the developmental potential
of aneuploid mouse embryos i.e. those with one chromosome extra (trisomics) or missing (monosomics). Death of monosomics preceeds those of the
comparable trisomics. For different chromosomes, trisomic and monosomic
developmental endpoints do overlap. Monosomic mouse embryos can die from
about the 8-16 cell stages on. Bearing in mind that monosomic and trisomic
embryos are both generated through the mechanism of meiotic non-disjunction, and that their frequencies of origin are identical, the data presented in Table 1 are sufficient proof of the relative absence of aneuploidy as a cause of genetic embryonic death in the pig.

Triploidy in the mouse leads to embryonic death during organogenesis
(Wroblewska, 1971) and in man, a similar embryonic stage is reached (Boué
et al., 1975). Table 1 gives some evidence for the occurrence of polyploidy, mainly triploidy in pig preimplantation embryos. In view of the earlier results of Hunter (1967), Hunter and Léglise (1971) and Hunter
(1972) concerning the susceptibility of the pig secondary oocyte to polyspermic fertilization after respectively, postovulatory oocyte aging and
with high numerical sperm/egg ratios around the time of fertilization,
this finding at first sight is not surprising.

However, Hunter and Léglise (1971) and Hunter (1972) present evidence that supernumerary sperm heads decondense within the oocyte but do not engage in syngamy. In man, both dispermy (5 cases) and fertilization by a diploid spermatozoon (2 cases after a second meiotic division reconstitution event) have been proven to yield triploid abortuses. (Kajii and Niikawa, 1977 using a Q and R-fluoresence chromosome banding technique.) In the mouse, both a dispermic fertilization and (more rarely) the fertilization by a diploid spermatozoon lead to chromosomes with a first cleavage morphology (Fraser and Maudlin, 1979). For the pig the quantitative relation between polyspermic fertilization and polyploidy, notably triploidy should not be difficult to assess when the procedures outlined by Hunter (1967, 1972) and Hunter and Léglise (1971) are combined with chromosome analysis of cleavage stages or expanded blastocysts.

THE OCCURRENCE OF CHROMOSOME ABNORMALITIES AMONG EMBRYOS OF OTHER MAMMALIAN SPECIES INCLUDING MAN

Table 2 gives incidences of aneuploidy and polyploidy for a number of species. Three observations are worth commenting on. Besides the human, the sheep is the only species discovered up to now with a considerable level of aneuploidy. In man, both aneuploidy and polyploidy are of considerable importance during early embryonic life. From the data presented in Table 2, their frequencies at conception cannot be accurately assessed. Another illustration of their frequent occurrence is their incidence among spontaneous abortuses (41.4% for aneuploidy and 12.2% for polyploidy, Boué et al. 1975).

The frequency of mixoploid cow blastocysts is remarkably high (41.5%). However, returning to Table 1, it can be seen that when in 10 days old pig blastocysts, trophoblast tissue was separated from inner cell mass tissue, polyploidy visible as multiplications of normal diploid (2n) cells preferentially occurred in the trophoblast cells. Polyploid cells were already encountered in some 4 days old pig blastocysts by van der Hoeven et al. (1985, in the press). The normal nature of polyploid trophoblast cells in mouse blastocysts was already demonstrated by Barlow et al. (1972).

THE RELATIVE IMPORTANCE OF CHROMOSOME ABNORMALITIES TO EMBRYONIC DEATH IN
MOUSE, MAN AND PIG

Due to the genetic differences that occur among the various mouse inbred and random-bred lines, it is difficult to give one figure for the
total level of prenatal death in the mouse. By experience, it can be
stated to vary around 10%. Generally, the level of prenatal death is regarded to be low in laboratory rodents. Comparing with Table 2, $\sim$ 30% of
this amount is due to chromosome abnormalities. In man, the level of total
embryonic wastage can only be guessed at. Ford (1975) states that "the proportion of human zygotes that are chromosomally abnormal could be as high
as 20%". The level of spontaneous abortions as a percentage of recogniz-
able pregnancies generally is taken to be in-between 15 and 20% (see Ford,
1975 as well). A statement that about half of the high prenatal death rate
in man is due to chromosome abnormalities is no more than an educated
guess.

Many reports give data about the embryonic death rate in the pig of
which a range of 30-40%, on the basis of the counts of corpora lutea and
assuming nearly complete fertilization, seems to be a fair assessment (see
for instance Perry and Rowlands, 1962). When returning to Table 1 it can
be seen that chromosome abnormalities never can contribute greatly to this
figure. Especially when insemination always preceeds ovulation, as in the
study of Dolch and Chrisman (1981), polyploidy is absent and consequently,
as aneuploidy has not yet been found among young pig embryos, there are no
chromosomal causes of embryonic death in the pig.

This project was carried out as part of the research of the working
party "Early pregnancy" from the Agricultural University, Wageningen,
The Netherlands.

REFERENCES

Akesson, A. and Henricson, B. 1972. Embryonic death in pigs caused by un-
    balanced karyotype. Acta vet. scand., 13, 151-160.
Barlow, P.W. and Sherman, M.I. 1972. The biochemistry of differentiation
    of mouse trophoblast: studies on polyploidy. J. Embryol. exp. Morph.,
    27, 447-465.
Binkert, F. and Schmid, W. 1977. Pre-implantation embryos of chinese
    hamster. I. Incidence of karyotype abnormalities in 226 control em-
    bryos. Mutation Res., 46, 63-76.
Boer, P. de and de Maar, P.H.M.D. 1976. A histological study of embryonic
    death caused by heterozygosity for the T26H reciprocal mouse translo-
    cation. J. Embryol. exp. Morph., 35, 595-606.

Bond, D.J. and Chandley, A.C. 1983. Aneuploidy. Oxford monographs on medical genetics no. 11. Oxford Univ. Press.

Boué, J., Boué, A. and Lazar, P. 1975. The epidemiology of human spontaneous abortions with chromosomal anomalies. In "Aging gametes, their biology and pathology" (Ed. R.J. Blandau) (S. Karger, Basel etc.) pp. 330–348.

Dolch, K.M. and Chrisman, C.L. 1981. Cytogenetic analysis of preimplantation blastocysts from prepuberal gilts treated with gonadotropins. Am. J. Vet. Res., 42, 344–346.

Feichheimer, N.S. and Beatty, R.A. 1974. Chromosomal abnormalities and sex ratio in rabbit blastocysts. J. Reprod. Fert., 37, 331–341.

Ford, C.E. 1975. The time in development at which gross genome imbalance is expressed. In: "The early development of mammals" (Ed. M. Balls and A.E. Wild), (Brit. Soc. for Dev. Biol. Symp. 2. Cambridge University Press). pp. 285–304.

Fraser, L.R. and Maudlin, I. 1979. Analysis of aneuploidy in first cleavage mouse embryos fertilized in vitro and in vivo. Envir. Hlth. Perspect., 31, 141–150.

Gropp, A. 1981. Chromosomenaberrationen, Geschwülste und Entwicklungsstörungen. Klin. Workenschr., 59, 965–975.

Hare, W.C.D., Singh, E.L., Betteridge, K.J., Eaglesome, M.D., Randall, G.C.B., Mitchell, D., Bilton, R.J. and Trounson, A.O. 1980. Chromosomal analysis of 159 bovine embryos collected 12 to 18 days after estrus. Can. J. Genet. Cytol. 22, 615–626.

Hoeven, F.A. van der, Cuijpers, M.P. and de Boer, P. 1985. Karyotypes of three or four days old pig embryos after short in vitro culture. J. Reprod. Fert. (in the press).

Hunter, R.H.F. 1967. The effects of delayed insemination on fertilization and early cleavage in the pig. J. Reprod. Fert., 13, 133–147.

Hunter, R.H.F. 1972. Local action of progesterone leading to polyspermic fertilization in pigs. J. Reprod. Fert., 31, 433–444.

Hunter, R.H.F. and Léglise, P.C. 1971. Polyspermic fertilization following tubal surgery in pigs, with particular reference to the role of the isthmus. J. Reprod. Fert., 24, 233–246.

Kajii, T. and Mikamo, K. 1978. Anatomic and chromosomal anomalies in 944 induced abortuses. Hum. Genet., 43, 247–258.

Kajii, T. and Niikawa, N. 1977. Origin of triploidy and tetraploidy in man: 11 cases with chromosome markers. Cytogenet. Cell Genet., 18, 109–125.

King, W.A., Gustavsson, I., Popescu, C.P. and Linares, T. 1981. Gametic products transmitted by rcp(13q–;14q+) translocation heterozygous pigs, and resulting embryo loss. Hereditas, 95, 239–246.

Long, S.E. and Williams, C.V. 1980. Frequency of chromosomal abnormalities in early embryos of the domestic sheep (Ovis aries). J. Reprod. Fert., 58, 197–201.

Long, S.E. and Williams, C.V. 1982. A comparison of the chromosome complement of inner cell mass and trophoblast cells in day-10 pig embryos. J. Reprod. Fert., 66, 645–648.

Lüning, K.G. 1975. Spontaneous recessive lethal mutation in the mouse. Mutation Res., 27, 367–373.

McFeely, R.A. 1967. Chromosome abnormalities in early embryos of the pig. J. Reprod. Fert., 13, 579–581.

Perry, J.S. and Rowlands, I.W. 1962. Early pregnancy in the pig. J. Reprod. Fert., 4, 175–188.

Popescu, C.P. and Boscher, J. 1982. Cytogenetics of preimplantation em-
    bryos produced by pigs heterozygous for the reciprocal translocation
    (4q+;14q-). Cytogenet. Cell Genet., 34, 119-123.
Smith, J.H. and Marlowe, T.J. 1971. A chromosomal analysis of 25-day-old
    pig embryos. Cytogenetics, 10, 385-391.
Wroblewska, J. 1971. Developmental anomaly in the mouse associated with
    triploidy. Cytogenetics, 10, 199-207.

214

TABLE 1 Chromosome studies of pre- and post implantation pig embryos as reported in the literature and by the present authors. Percentages of chromosome abnormalities are given between parentheses. n = ploidy-level.

Reference	Race	Embryonic (days)	N age	Aneu-ploidy	Poly-ploidy	Mixo-ploidy	Other
McFeely (1967)	variable	10	86	-	7(8.0) 3n	1(1.1)	1(1.1)
Dolch & Chrisman (1981)*	Yorkshire x (Yorkshire x Hampshire)	10	169	-	-	-	-
Long & Williams (1982)	Large white	10	38 (trophoblast)	-	1(2.6) 3n/6n	17(44.7)**	-
			39 (inner cell mass)	-	2(5.1) 3n	2(5.1)	-
Van der Hoeven et al. (1985)	Dutch landrace	4	82	-	1(1.2) n	1(1.2)	-
Smith & Marlowe (1971)	--		68	1(1.5)†	-	-	-

* Embryos recovered after superovulation
** Mixoploids explainable through genome duplication in trophoblast cells
† A 2n/2n-1 mosaic

TABLE 2 Chromosome studies of embryos of various species.

Reference	Species	Stage	N	Aneu-ploidy (%)	Poly-ploidy (%)	Mixoploidy (%)
Fraser & Maudlin (1979)	mouse	1 cell	840	1.4%	1.8% (n=228)	-
Binkert & Schmid (1977)	chinese hamster	4-8 cells	226	0.9%	3.1%	-
Fechheimer & Beatty (1974)	rabbit	5-day blastocyst	463	0.9%	1.7%	1.3%
Long & Williams (1980)	sheep	2-8 cells	89	4.7%	-	1.1%
Hare et al. (1980)	cattle	12-15 day blastocyst	159	-	0.6%	1.2% 41.5%*
Martin cit. by Bond & Chandley (1983)	human	spermatozoa	948	5.2%	?	-
Kajii and Mikamo † (1978)	human	induced abortions (4-14 weeks)	728	1.8%	1.0%	-

* Mixoploids that are based on polyploid trophoblast cells.
† Frequency strongly dependent on embryonic age

216

CYTOLOGICAL PARAMETERS FOR RATING BOVINE EMBRYO QUALITY

B.M.J.L. Mannaerts

Department of Functional Morphology
Faculty of Veterinary Sciences
State University
Yalelaan 1, 3508 TD Utrecht, The Netherlands

ABSTRACT

Cytological studies in addition to morphological examination of bovine embryos may result in a better estimation of embryo quality. In day 7 embryos of good quality the total cell number, the mitotic index, the karyotype, and the DNA contents of the interphase nuclei were studied. The value of the study of these parameters for rating embryo quality is discussed.

INTRODUCTION

In connection with the commercial applications of embryo transfer there is need for reliable methods of rating embryo quality. As the evaluation of embryos based on morphological characteristics remains subjective, the study of cytological parameters may result in a better estimation of embryo quality. By establishing the total cell number and the mitotic index, more information can be obtained about the stage of development and the mitotic activity of the embryo (King et al., 1979). Karyotype analysis before embryo transfer may provide opportunities for sex determination (Singh and Hare, 1980), and for the detection of chromosomal abnormalities (King et al., 1981; Hare et al., 1980). The study of DNA contents of interphase nuclei may lead to additional information about the mitotic activity, and to the identification of polyploid embryos.

In this paper preliminary results of cytological studies on morphologically normal day 7 embryos are presented.

MATERIALS AND METHODS

Embryo Collection and Culture

Morula to blastocyst stage bovine embryos were collected from superovulated animals, by a non-surgical method, 7 days after insemination. The embryos were classified according to the criteria of Lindner and Wright (1983). Embryos of good quality were selected for cytological studies. They were handled in Dulbecco's phosphate-buffered saline (PBS), supplemented as described by Whittingham (1971). For culturing embryos, heat-inactivated fetal calf serum (10-20 %) was added to this medium. Embryos were incubated for 20 to 24 hrs, at 37 °C. To obtain mitotic arrest, either colchicine (0.5 µg/ml medium) was used during the last 3 to 4 hrs of the culture period, or vinblastine (0.005 µg/ml medium) during the whole culture period.

Microscopic Preparations

Preparations were made from entire embryos, directly after collection or after culturing. To harvest, the embryos were placed in a hypotonic solution (0.075 M potassium chloride) for 10 to 15 min. The procedure followed thereafter was similar to the method for making chromosome preparations from bovine zygotes and blastocysts as described by King et al. (1979). This included fixation in a mixture of acetic acid and methanol (1:1, by volume). A second fixation was carried out by placing the preparations in a freshly prepared mixture of absolute methanol, formaldehyde (35 %) and glacial acetic acid (85:10:5 by volume) for 60 min. This second fixation as well as all following treatments were performed at room temperature (20-22 °C). After fixation the preparations were washed in 0.1 M citric acid/citrate buffer (pH 5.6) and in distilled water, and were air-dried.

Feulgen-Staining and Stage-Scanning Cytophotometry

Preparations were hydrolyzed in 5 N HCl for 1 hr, and stained with a pararosaniline(SO_2) Schiff medium prepared according to Graumann (1953) (60 min.). The excess of Schiff reagent was removed by washing in sulfurous acid (3 x 10 min.), in citric acid/citrate buffer (pH 5.6) (Duijndam and Van Duijn, 1973), and in distilled water. Dehydration was obtained via a graded alcohol-xylene series, and preparations were embedded in Fluoromount (Gurr). Absorbance scanning was performed at 559 nm with a Cytoscan SMP (Zeiss, Oberkochen, FRG) interfaced to a PDP 11/10 computer with a HIDACSYS-ARRAYSCAN program (Van der Ploeg et al., 1977). Measurements were taken at 0.5 µm intervals. For each interphase nucleus the integrated absorbance (obtained by summation of the local absorbance values after subtraction of the mean background value) was stored together with the number of object points.

RESULTS

Microscopic preparations were made from 36 embryos. The total cell number and the mitotic index were determined by counting the interphase nuclei and the metaphase spreads. The mean number of cells in embryos of the morula/early blastocyst stage, the blastocyst stage, and the expanded blastocyst stage were 98.2 (range 72-146), 122.3 (range 99-172), and 163.2 (range 105-261), respectively. In 14 embryos the mitotic index was determined directly after collection, in 9 embryos after a culture period of 20 to 24 hours (Table 1). All uncultured embryos, with the exception of one, showed cells in mitosis (1.9 % ± 1.4, mean mitotic index ± SD). However, the mitotic activity of the cultured embryos appeared to be considerably higher (5.1 % ± 2.7). In other experiments mitotic inhibitors were used to accumulate cells in mitosis. Six embryos were cultured in the presence of colchicine, 7 embryos in the presence of vinblastine. It appeared that by colchicine treatment the mean mitotic index was only slightly increased (6.1 % ± 2.0), whereas by by application of vinblastine the mean mitotic index was largely increased (15.6 % ± 10.6).

TABLE 1 The number of embryos investigated, and their mitotic indices.

	No. of embryos	Mitotic index (Mean ± SD)
Without culture	14	1.9 % ± 1.4
Cultured 20-24 hrs	9	5.1 % ± 2.7
Cultured 20-24 hrs, final 3-4 hrs with colchicine	6	6.1 % ± 2.0
Cultured 20-24 hrs, with vinblastine	7	15.6 % ± 10.6

The metaphase spreads obtained were usually of insufficient quality for complete karyotype analysis, but often of sufficient quality for sex determination and for the detection of certain chromosome abnormalities, such as polyploidy and translocations of the centric fusion type. However, in the studied material these abnormalities were not observed.

With the Feulgen staining method not only interphase nuclei, but also
chromosomes stained rather well. So this method enabled the study of
different cytological parameters in the same preparation. The DNA contents
of all interphase nuclei were determined in 5 cultured and 3 uncultured
embryos. The frequency distribution of the DNA absorbance values of the
nuclei of an uncultured embryo is shown in Fig. 1.

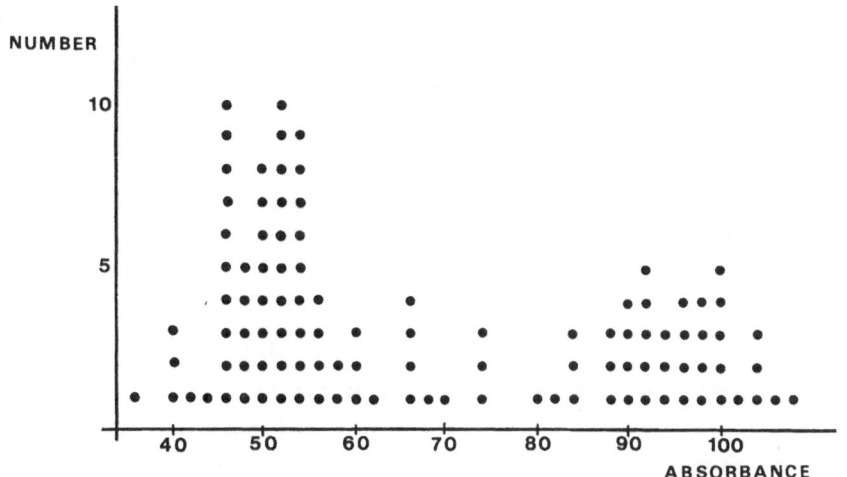

Fig. 1   Frequency distribution of the DNA absorbance values of the
nuclei of an uncultured embryo (mitotic index 0 %).

Nuclei with low absorbance values may be considered to belong to
cells in the G1 phase, while absorbance values about twice as high come
either from cells in the G2 phase, or from cells in the late-S phase. The
few nuclei with intermediate absorbance values probably belong to cells in
the mid-S phase.

Fig. 2 shows the frequency of distribution of the DNA absorbance
values of the nuclei of a cultured embryo. The differences between the
frequency distributions shown in Figs. 1 and 2 suggest that the mitotic
activity of the cultured embryo was higher than that of the uncultured
embryo, as it has relatively less cells in the G1 phase and more cells in
the S phase and G2 phase.

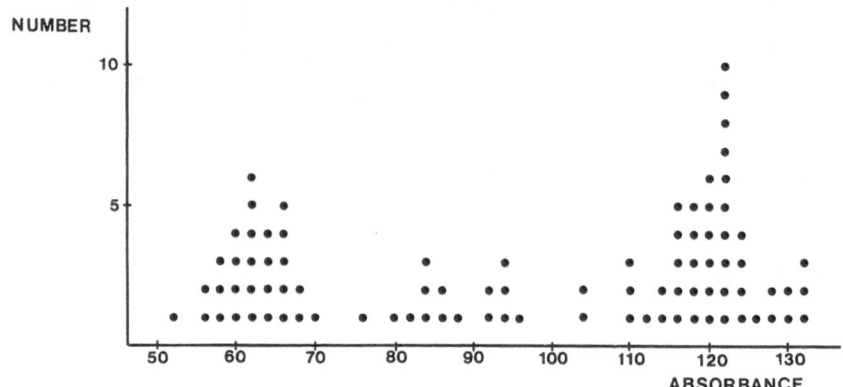

Fig. 2   Frequency distribution of the DNA absorbance values of the
nuclei of a cultured embryo (mitotic index 7.7 %).

DISCUSSION

In the present study, embryos of the same developmental stage generally
had an appropriate mean number of cells. Earlier studies have shown that
day 7 embryos of repeat-breeder heifers have a reduced mean number of
cells in comparison with day 7 embryos from virgin heifers (King and
Linares, 1983). As a reduced total cell number may be an indication of
cell death or of a low mitotic activity, embryos with a relatively low
number of cells might be less viable than embryos with a relatively high
cell number.

Preparations made from embryos directly after collection showed a
considerably lower mean mitotic index than preparations made from embryos
cultured for 20 to 24 hours. This result is in good agreement with the
idea that embryos require culture for at least some hours, in order to
recover from the trauma induced by collection (King et al., 1979). The
mean mitotic index was only slightly increased by colchicine treatment,
but largely increased by application of vinblastine, since vinblastine was
able to arrest dividing cells during the whole culture period, while
colchicine could only arrest dividing cells during the last 3 to 4 hours
of the culture period. So the use of vinblastine may provide opportunities
to increase the number of metaphase spreads, which is particularly useful
when studying embryos with a low mitotic activity.

Although spontaneous chromosome abnormalities are probably rare in embryos from karyotypically normal cattle, it seems important to screen embryos for certain chromosome abnormalities. Various studies in sheep (Williams and Long, 1980), rabbit (Fujimoto et al., 1974), and cattle (King et al., 1979; King et al., 1981) have indicated that superovulation may increase the incidence of polyspermic fertilization and retained mitotic activity of the polar body. This may result in abnormalities such as haploidy, triploidy and mixoploidy. In the present study, which included only embryos of good quality, chromosome abnormalities were not observed. However, there are indications that chromosome abnormalities are more prevalent in day 7 embryos of fair quality (King, 1983).

Karyotype screening of embryos is also important because in a large number of cattle breeds animals occur that are phenotypically normal but karyotypically abnormal, in that chromosome no. 1 and chromosome no. 29 are fused. Animals, both male and female, that are heterozygous for this so-called 1/29 translocation have a reduced fertility (Gustavsson, 1980). This is believed to be due to the formation, and subsequent death, of chromosomally unbalanced embryos (King et al., 1981). On the other hand, embryos with a balanced karyotype, but carrier of the translocation, will cause spreading of the abnormality. In the future it will be advantageous to combine chromosome examination and embryo transfer, as this will make it possible to select for sex as well as for a normal karyotype.

The results of the DNA measurements indicate that the study of DNA contents of interphase nuclei provides a good alternative method of rating embryo quality. As the distribution of the absorbance values reflects the mitotic activity, it may give additional information about the viability of the embryo. Further studies will be needed to standardize the Feulgen staining procedure, in order to make the study of small numbers of cells reliable.

The study of cytological parameters in embryos will become useful in practice in combination with the possibility to split embryos. Cytological studies could be carried out on parts of embryos, while the other parts would remain available for transfer.

ACKNOWLEDGEMENTS

This work was supported by a grant from the Royal Dutch Cattle Syndicate.

I am greatly indebted to Mr. H. Hoogenkamp and Mr. H. Heuveling of the Clinic of Veterinary Obstetrics, Gynaecology and A.I. for collecting the embryos. Sincere thanks are also due to the Department of Histochemistry and Cytochemistry (State University, Leiden), especially to Dr. W.A.L. Duijndam, for making available the facilities for the DNA measurements, and for their stimulating interest. I am grateful to Dr. A.A. Bosma for helpful suggestions, and for critical reading of the manuscript.

REFERENCES

Duijndam, W.A.L. and Van Duijn, P. 1973. The dependance of the absorbance of the final chromophore formed in the Feulgen-Schiff reaction on the pH of the medium. Histochemie, 35, 373-375.

Fujimoto, S., Pahlavan, N. and Duklow, W.R. 1974. Chromosome abnormalities in rabbit preimplantation blastocysts induced by superovulation. J. Repr. Fert., 40, 177-181.

Graumann, W. 1953. Zur Standardisierung des Schiffschen Reagens. Z. Wiss. Mikrosk., 61, 225-226.

Gustavsson, I. 1980. Chromosome aberrations and their influence on the reproductive performance of domestic animals - a review. Z. Tierzüchtg. Züchtgsbiol. 97, 176-195.

Hare, W.C.D., Sing, E.L., Betteridge, K.J., Eaglesome, M.D., Randall, G.C.B., Mitchel, D., Bilton, R.J. and Trounson, A.O. 1980. Chromosomal analysis of 159 bovine embryos collected 12 to 18 days after estrus. Can. J. Genet. Cytol., 11, 287-293.

King, W.A., Linares, T., Gustavsson, I. and Bane, A. 1979. A method for preparation of chromosomes from bovine zygotes and blastocysts. Vet. Sci. Comm., 3, 51-56.

King, W.A., Linares, T. and Gustavsson, I. 1981. Cytogenetics of preimplantation embryos sired by bulls heterozygous for the 1/29 translocation. Hereditas, 94, 219-224.

King, W.A. and Linares, T. 1983. A cytogenetic study of repeat-breeders and their embryos. Can. Vet. J., 24, 112-115.

King, W.A. 1983. Cytogenetics of embryos: application to the livestock industry. II Symposium on Advanced Topics in Animal Reproduction, Brazil, August 8-11, pp. 113-132.

Lindner, G.M. and Wright, R.W. 1983. Bovine embryo morphology and evaluation. Theriogenology, 20, 407-416.

Singh, E.L. and Hare, W.C.D. 1980. The feasibility of sexing bovine morula stage embryos prior to embryo transfer. Theriogenology, 14, 421-427.

Van der Ploeg, M., Van den Broek, K., Smeulders, A.W.M., Vossepoel, A.M. and Van Duijn, P. 1977. HIDACSYS: Computer programs for interactive scanning cytophotometry. Histochemistry, 54, 273-288.

Whittingham, D.G. 1971. Survival of mouse embryos after freezing and thawing. Nature, 233, 125-126.

Williams, C.V. and Long, S.E. 1980. The effect of superovulation on the chromosome complement of early sheep embryos. 4th Europ. Colloq. Cytogenet. Domest. Anim., Uppsala, pp 168-171.

EMBRYONIC LOSSES AND THE ROLE OF NUTRITION
- STATISTICAL CONSIDERATIONS

N.O. Rasbech
Institute for Animal Reproduction
The Royal Veterinary and Agricultural University
Bülowsvej 13, DK-1870 Copenhagen V, Denmark

ABSTRACT

Data on correlation between nutritional regimes and proven embryonic losses in cattle and swine are negligible. Widespread inconsistency seems to exist concerning criteria or terms to categorize experimental animals subjected to investigations for embryonic losses. For this reason the author has felt it necessary to discuss some highly variable reproductive parameters found in any farm animal population and which may be of importance for evaluation of embryo losses in general and for designing research projects aiming at estimation of correlation between nutrition and proven embryonic losses in particular. At least four interrelated and highly variable parameters of importance for the reproductive efficiency in farm animal herds (dairy cattle) should be considered 1) effectiveness of inseminations (CR), 2) heat detection rate (HDR), 3) days from calving to 1st ins. (C-ins.), and 4) deadline for insemination of cows post partum (see Table 5). From field investigations it appears that in a normal cattle population between 50-60% of the cows may be expected to concieve after an insemination irrespectively whether it is after 1st, 2nd, 3rd, 4th, or 5th successive insemination (Fig. 1) and that regular/irregular interservice intervals occur in a normal population after a variable pattern (mainly due to external factors)(Table 6). From a statistical point of view it is possible to calculate the number of animals required in experimental and control groups to obtain significant differences (Tables 7 and 8). It seems likely that such experiments would require large groups of animals and might bring additional variable factors causing conflicting results.

INTRODUCTION

According to the Committee for Reproductive Nomenclature (1972) embryonic mortality refers to fertility losses during the embryonic period from conception to completion of the stage of differentiation, which in the cow occurs approximately at 45 days of gestation. It is generally accepted that pregnancy rates or non-return rates reflect the percentage of developing embryos after a fixed deadline during the gestation period and include fertilization failures as well as losses of embryos after fertilization. Research projects designed to throw light on factors affecting the embryonic losses should therefore, strictly speaking, only provide data on losses of established embryos during the embryonic period and conclusions drawn from such data should be based on statistically significant differences between control and experimental groups of animals. Such a procedure will necessitate

to distinguish between fertilizing failures and embryonic losses and will
usually require planned slaughter.

Over the years many reviews and research papers have been published
(see review of Ayalon 1978) dealing with embryonic mortality in general in
farm animals. A majority of authors base their observations on data which
include fertilization failures, while others have made an attempt to provide data for fertilization failures usually by planned slaughter on day 3
after mating and data for subsequent embryonic losses at different times
usually up to 35 days after mating.

The experimental animals used in such experiments have usually been
selected after criteria such as 1) gynaecologically normal heifers or cows
with regular/irregular cycles, and 2) repeat breeder cows which have failed
to conceive after 3 or 4 inseminations followed by regular/irregular interservice periods. The often conflicting data reported could be due to difficulties to categorize the experimental animals under terms such as repeat
breeders or animals with irregular interservice periods or due to a too low
number of animals in control and experimental groups to obtain statistical
significant differences between groups.

There are numerous publications dealing with the role of nutritional
factors (high and low plane of nutrition, addition of vitamins and minerals
a.o.) based on conception rates, return rates and/or progesterone profiles
in cattle as well as in swine. Such data, however, include losses from fertilizing failures as well as embryonic losses.

The role of nutrition on embryonic losses

There are practically no experimental data on the correlation between
nutrition and proven embryonic mortality in cattle. In one experiment HILL
et al. (1970) compared data obtained from a group of 20 undernourished beef
heifers with data obtained from 20 control heifers. Both groups were fed
with the same diet, but the undernourished heifers received a lower amount
of the diet representing 83% of the maintenance requirements for energy and
protein.

The conclusion was drawn that undernourishment reduced the plasma levels of progesterone within 5 days and reduced the proportion of heifers
with normal fertilized ova. Weights of corpora lutea in the undernourished
group were reduced by 75% of the control values. No clear effect was observed on embryonic mortality 8 or 18 days after breeding (see Table 1).

TABLE 1. Influence of nutritional regimes on fertility (Hill et al. 1970)

	3 days			8 days			18 days		Overall fertility	
		Ova	Ova		Ova	Ova		No.		No.
	No.	recov-	fertil-	No.	recov-	fertil-	No.	preg-	No.	preg-
Treatment	heifers	ered	ized	heifers	ered	ized	heifers	nant	heifers	nant
Control	5	5	5	4[a]	3	3	5	3	14	11
Undernutrition	5[b]	5	3[c]	5	2	2	5	2	15	6

Time of recovery after mating

[a] One heifer deleted because the oviduct was damaged and recovery impossible
[b] Two additional heifers were studied in this group but ova were not recovered
[c] One fertilized ovum was considered abnormal on the basis of irregular cell size and granular degeneration

In swine a considerable number of publications are dealing with the nutritional role on the breeding efficiency based on differences in sexual maturity, litter size, and survival of piglets after weaning. It has only been possible to find one publication (Koefoed-Johnsen et al. 1955)where planned slaughter on day 4-6 and 4 weeks after mating has been performed to provide data on the fertilization rate (no. of ovulations) and the number of established embryos. The procedure and main results are cited below in Table 2.

TABLE 2. The importance of vitamin B_{12} on the fertility in sows and gilts (Koefoed-Johnsen et al. 1955).

	I	II	III
	20 gilts	20 gilts	20 gilts
	Basic diet + 2-3 kg skimmilk	Basic diet $-B_{12}$	Basic diet + 80 ɣ B_{12}
Slaughter on day 4-6	10 gilts	10 gilts	10 gilts
Weight of uterus (g)	413	386	356
Weight of ovaries (g)	11.8	10.9	13.0
No. of ovulations	16.1	12.3	11.7
Slaughter, 4 weeks	10 gilts	10 gilts	10 gilts
No. of fetuses	9.9	5.6	10.4
Weight of fetuses (g)	1.87	1.32	1.35
Weight of ovaries (g)	15.6	12.7	13.3

The conclusions drawn from these observations were that the group which did not receive animal protein or vitamin B_{12} supplement had lower weight of fetuses when slaughtered 4 weeks after mating compared with the control group. The ovulation rate was highest in the group which received skimmilk.

When discussing the very scanty information on the role of nutrition on embryonic survival the question arises whether it is possible or necessary to distinguish between fertilization rates and embryonic losses when experimental short or long term feeding experiments are designed because the effect of nutritional factors may inevitably affect not only the embryonic survival rate but also the fertilization rate.

Estimation of embryonic losses from a statistical point of view

Since it is practically impossible to find experimental data on the correlation between nutrition and proven embryonic mortality and since there seems to exist widespred inconsistency concerning criteria or terms (such as the repeat breeder term, irregular interservice intervals and chances for obtaining pregnancy after 1st, 2nd, 3rd, 4th, 5th, or 6th insemination a.o.) to categorize experimental animals subjected to investigations for embryonic losses the author has felt it necessary to discuss some highly variable reproductive parameters found in any farm animal population and which may be of importance for the estimation of embryonic losses in general and for designing research programmes aiming at an estimation of the correlation between nutrition and proven embryonic losses in particular.

Variable parameters (factors) for evaluation of reproductive efficiency in farm animals (dairy herds)

At least four variable parameters seem to be important for the evaluation of the breeding efficiency in a herd (Pedersen 1982):

1) Effectiveness of inseminations = conception rate (CR)
2) Heat detection rate (HDR)
3) Days from calving to 1st ins. in days (C-ins.)
4) Deadline for inseminations of cows post partum

Effectiveness of ins. = % pregnancies/insemination = conception rate (CR)

If we assume that the effectiveness of inseminations in a cow population is constant and independent of whether inseminations of cows are continued for a longer or shorter period, then in order to obtain an aimed pregnancy percentage it will be necessary to continue insemination for a longer period if the effectiveness of inseminations is low than it would be if the effectiveness is high. This can be illustrated in Table 3 by variable effectiveness.

TABLE 3. Percent pregnant cows with successive inseminations of varying effectiveness (Rottensten 1948).

Ins.	Effectiveness				
	30%	40%	50%	60%	70%
1.	30.0	40.0	50.0	60.0	70.0
2.	51.0	64.0	75.0	84.0	91.0
3.	65.7	78.4	87.5	93.6	97.3
4.	76.0	87.0	93.8	97.4	99.2
5.	83.2	92.2	96.9	99.0	99.8
6.	88.2	95.3	98.4	99.6	99.9
7.	91.8	97.2	99.2	99.8	-
8.	94.3	98.3	99.6	99.9	-
9.	96.0	99.0	99.8	-	-
10.	97.2	99.4	99.9	-	-
11.	98.0	99.6	-	-	-
12.	98.6	99.8	-	-	-
13.	99.0	99.9	-	-	-
14.	99.3	-	-	-	-
15.	99.5	-	-	-	-
16.	99.7	-	-	-	-
17.	99.8	-	-	-	-
18.	99.9	-	-	-	-

If it is accepted that the effect of inseminations may vary from 30% to 70%, due to factors such as semen quality of low fertilizing capacity, contaminated semen, technical errors, failures in timing inseminations a.o., it is possible to calculate the average no. of inseminations which are needed to obtain an aimed percentage (illustrated in Table 4). E.g. to obtain a pregnancy percentage of 85, 1.31 inseminations per pregnant cow are required at an effectiveness of 60%, but 2.4 inseminations are required at an effectiveness of 30%.

TABLE 4. Number of inseminations with different pregnancy percentage and effectiveness (Rottensten 1948).

Effectiveness %	Total ins.* per cow	Number of inseminations per pregnant cow with conception percentage						
		65	70	75	80	85	90	95
70	1.43		1.00	1.07	1.12	1.18	1.22	1.31
65	1.54	1.00	1.07	1.13	1.19	1.24	1.30	1.42
60	1.67	1.08	1.14	1.20	1.25	1.31	1.40	1.49
55	1.82	1.15	1.21	1.27	1.31	1.41	1.50	1.62
50	2.00	1.23	1.29	1.33	1.44	1.53	1.63	1.77
45	2.22	1.31	1.36	1.47	1.56	1.66	1.79	1.95
40	2.50	1.40	1.51	1.61	1.72	1.85	2.01	2.19
35	2.86	1.58	1.67	1.79	1.92	2.08	2.27	2.48
30	3.33	1.75	1.90	2.04	2.21	2.40	2.62	2.89

* When all cows are pregnant

The figures in tables 3 and 4 are based on the assumption that effectiveness is the same whether 1, 2, or more inseminations are required. It may be discussed how close this assumption is fullfilled under practical condi

228

tions. However, experiences from a number of calculations made on cattle populations within field insemination centres do indicate that the average chances for pregnancy in large cattle populations are insignificantly different whether one or more inseminations are used under the assumption that the animals are not subjected to clinical detectible gynaecological pathological conditions. This was pointed out as early as in 1948 in Denmark by Rottensten and later confirmed by Hoppe (1954) and Hyttel (1982).

Over a 10 year period Hoppe investigated in an AI association the percentage of cows which concieved after 1st, 2nd, 3rd, 4th, or later inseminations out of a total number of 146639 pregnant cows. Results are shown in Fig. 1.

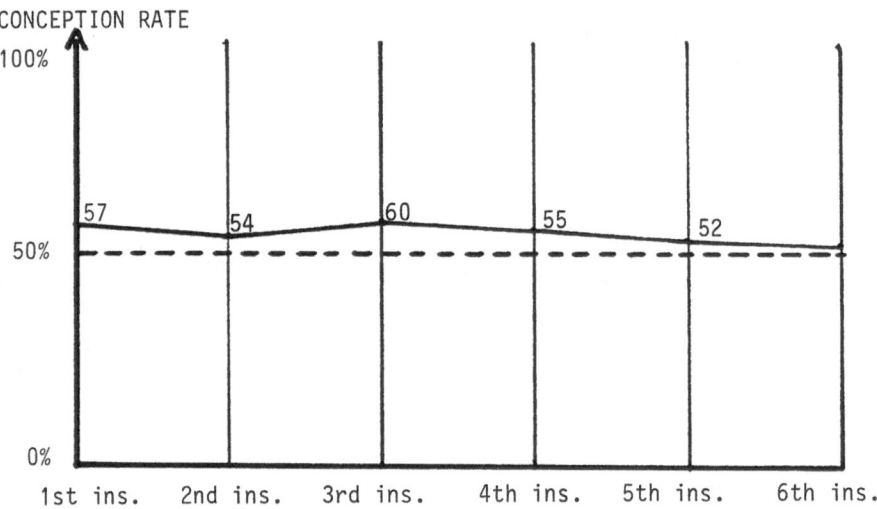

FIGURE 1. Conception rates (%) after 1st, 2nd, 3rd, 4th or 5th successive inseminations based on the total number of pregnant cows (146639 pregnant cows) during a 10 year period in an AI district (Hoppe 1954).

From these results it seems justified to draw the conclusion that out of an arbitrarily selected large cow population, between 50-60% will be pregnant after a single insemination irrespective of whether they have been inseminated 1, 2, 3, 4, or 5 times. For this reason the term repeat breeder cows is not accepted as a criteria or diagnosis for reproductive efficiency in individual cows under Danish conditions.

Heat detection rate (HDR) = the percentage of heat periods detected in a herd out of the total number of possible heat periods within a certain period.

From farm to farm the heat detection rates vary considerably for several reasons (from about 30-90%). About 90% is considered to be the highest
number of heat periods observed out of the total number of possible heat
periods in a herd due to anestrus post partum, very short heat periods a.o.
Since it is possible and important to calculate the efficiency of heat detection in traditional dairy herds we prefer to use the term heat detection
rate instead of using terms such as silent heat, subestrus, a.o..

Days from calving to 1st insemination (C-ins.)

This parameter varies considerably from farm to farm since it is dependent on whether a farmer aims to start inseminations of cows earlier or
later in the post partum period (i.e. on day 42, 63, or 84 p.p.).

Deadline for insemination of cows post partum

This parameter is highly dependent on the culling policy accepted for
the herd and may vary from roughly 100-300 days p.p.

If it is anticipated that 250 days p.p. represent the average deadline
for inseminations (culling?) in traditional herds it is possible to calculate the reproductive efficiency which may be expected within 250 days after
calving in a herd. This is illustrated in the simulation model, Table 5.

This model has been proved under field conditions in several larger
dairy herds. The expected parameters have shown to be in accordance with
those found under practical farm conditions. It is of interest to note that
an increased heat detection rate will result in more inseminations/cow and
increase the % returning cows (4 ins. or more) against an increase in the
percentage of pregnant cows with a shorter calving-conception interval (ED).
On the other hand this seems paradoxical, but is due to the fact that the
time (deadline) available for breeding is limited. From this model (Table
5) it is possible, taking the variable parameters into consideration, to
predict rather accurately the number of cows which will require 4 or more
inseminations per cow in a population of gynaecological normal cows.

Interservice intervals and reproductive efficiency

Regular or irregular interservice intervals in individual cows are
often used as a criteria for reproductive normality versus abnormality in
experimental cows used for the study of embryonic losses. In the original
work of Tanabe & Casida (1949) on embryonic mortality in dairy cattle, cows
with a low fertility were selected on the basis of having a minimum of 4

TABLE 5. Reproductive efficiency expected within 250 days after calving by measuring the following parameters: 1) total no. of pregnant cows (% pregnant cows), 2) calving conception interval (ED)*, 3) no. of inseminations required per inseminated cow (AI/ins. cow), and 4) percentage of cows requiring 4 inseminations or more (% return 4+) as a function of the conception rate (CR), the heat detection rate (HDR), and the distance in days from calving to aimed 1st insemination (C-ins.) (Pedersen 1982).

Conception rate (CR) %	Heat detection rate (HDR)	90			60			30		
	Calving to ins. (C-ins.) (days)	42	63	84	42	63	84	42	63	84
70	% pregnant.	100	100	100	100	99	99	91	88	85
	Aver. ED	54	75	96	70	91	111	99	116	133
	AI/ins. cow	1,4	1,4	1,4	1,4	1,4	1,4	1,3	1,3	1,3
	% return (4+)	3	3	3	3	2	2	1	1	1
60	% pregnant.	100	100	100	99	98	97	86	83	80
	Aver. ED	60	81	102	77	97	116	104	121	137
	AI/ins. cow	1,7	1,7	1,7	1,7	1,6	1,6	1,5	1,5	1,4
	% return (4+)	6	6	6	6	6	5	2	2	1
50	% pregnant.	100	100	99	97	96	94	80	77	73
	Aver. ED	67	88	108	85	104	123	110	125	140
	AI/ins. cow	2,0	2,0	2,0	1,9	1,9	1,9	1,7	1,6	1,5
	% return (4+)	12	12	12	12	11	10	4	3	2
40	% pregnant	99	98	97	94	92	89	72	68	64
	Aver. ED	77	97	116	94	112	129	115	129	144
	AI/ins. cow	2,5	2,5	2,4	2,3	2,3	2,2	1,9	1,8	1,7
	% return (4+)	22	22	22	20	19	18	8	6	4
30	% pregnant.	96	94	92	86	83	80	61	57	53
	Aver. ED	89	108	126	104	121	137	120	134	147
	AI/ins. cow	3,2	3,1	3,1	2,9	2,8	2,7	2,1	2,0	1,9
	% return (4+)	34	34	34	32	31	28	12	9	7
20	% pregnant.	86	83	80	72	68	64	46	43	39
	Aver. ED	104	121	137	115	129	144	126	138	151
	AI/ins. cow	4,3	4,2	4,0	3,6	3,4	3,2	2,4	2,2	2,1
	% return (4+)	51	51	51	48	46	42	18	14	10
10	% pregnant.	61	57	53	46	43	39	26	24	22
	Aver. ED	120	134	147	126	138	151	131	143	154
	AI/ins. cow	6,1	5,7	5,3	4,6	4,3	3,9	2,7	2,5	2,3
	% return (4+)	73	73	73	69	66	60	26	20	24

* ED = empty days

services, being clinically normal on rectal palpation and having normal estrous cycles with normal interservice intervals between breedings.

In order to investigate, under Danish conditions, the frequency of regular/irregular interservice intervals in a normal cow population all intervals between 1st insemination and the following service were carefully calculated. The distribution of specified intervals is presented in Table 6 as the percentage of the total no. of interservice intervals.

TABLE 6. Distribution of interservice intervals (%) on total no. of intervals following 1st insemination in a cow population of 10690 cows (Rasbech 1981).

Interservice intervals	Distribution of intervals (%) on total no. of intervals
< 18 days	4 - 5
18 - 24 "	46 - 47
25 - 38	14 - 15
39 - 45	12 - 13
46 - 59	9 - 10
60 - 66	3 - 4
later than 66	7 - 8

The conclusion drawn from these observations was that regular/irregular interservice intervals occur after a rather variable pattern in a normal cow population due to a variety of external and biological factors (e.g. incorrect timing of ins., low detection rate, a.o.) and should not be used to categorize the reproductive efficiency of individual cows as being normal versus abnormal.

Statistical considerations concerning experimental animals and embryonic losses

In the following an attempt is made to estimate how many animals should be required in control and experimental groups respectively in order to find/detect by a given significance (i.e. $p < 0.05$) an existing (the hypothesis) difference in embryonic losses when a reasonable probability of success (i.e. 50%) is required. The calculation is made under the assumption that the above mentioned parameters are taken into account.

In order to obtain significant differences the number of animals required in each of the groups may be found from Table 7.

In Table 8 (example) it is anticipated (hypothesis) that the frequency of non pregnant animals at slaughter on day 3 (fertilizing failures) is 30% in experimental conditions and 25 in control conditions, and that the frequency on day 35 (fertility failure, embryonic losses) is 50% and 45% respectively. From Table 7 it is now possible to estimate the number of animals required in each group in order to be able to verify ($p < 0.05$) this hypothesis in at least 50% of the experiments carried out.

TABLE 7. Depending on the number of animals in each of two groups, the table give corresponding frequencies which are just different by statistical significance (p < 0.05, using χ^2-test with Yates correction (Pedersen 1984, unpublished).

Highest frequency (%)	Number of animals in each group																
	25	50	100	150	200	250	300	350	400	450	500	600	700	800	900	1000	5000
1	-	-	-	-	-	-	-	-	-	-	-	0.02	0.08	0.1	0.1	0.2	0.6
2	-	-	-	-	-	-	0.05	0.1	0.2	0.3	0.4	0.5	0.6	0.7	0.8	0.8	1.4
3	-	-	-	-	0.08	0.3	0.5	0.7	0.8	0.9	1.0	1.2	1.3	1.4	1.5	1.5	2.3
4	-	-	-	0.1	0.5	0.8	1.1	1.3	1.4	1.6	1.7	1.9	2.0	2.1	2.2	2.3	3.2
5	-	-	-	0.6	1.1	1.4	1.7	1.9	2.1	2.3	2.4	2.6	2.8	2.9	3.0	3.1	4.1
10	-	-	2.3	3.6	4.4	4.9	5.4	5.7	5.9	6.2	6.3	6.7	6.9	7.1	7.2	7.4	8.8
15	-	2.0	5.5	7.2	8.2	8.9	9.4	9.8	10.1	10.4	10.6	11.0	11.3	11.5	11.7	11.9	13.6
20	-	5.0	9.2	11.1	12.2	13.0	13.6	14.1	14.5	14.8	15.0	15.5	15.8	16.1	16.3	16.5	18.4
25	2.0	8.3	13.1	15.2	16.5	17.4	18.0	18.5	19.0	19.3	19.6	20.1	20.4	20.7	21.0	21.2	23.3
30	4.9	12.0	17.2	19.5	20.9	21.9	22.6	23.1	23.6	23.9	24.2	24.7	25.1	25.4	25.7	25.9	28.1
35	8.1	15.9	21.5	24.0	25.4	26.5	27.2	27.8	28.3	28.6	29.0	29.5	29.9	30.2	30.5	30.7	33.1
40	11.7	20.0	25.9	28.5	30.1	31.2	31.9	32.5	33.0	33.4	33.8	34.3	34.7	35.1	35.4	35.6	38.0
45	15.5	24.3	30.5	33.2	34.9	35.9	36.8	37.4	37.9	38.3	38.6	39.2	39.6	40.0	40.3	40.5	43.0
50	19.5	28.8	35.2	38.1	39.7	40.8	41.6	42.3	42.8	43.2	43.6	44.1	44.6	44.9	45.2	45.5	48.0
55	23.9	33.5	40.1	43.0	44.7	45.8	46.6	47.3	47.8	48.2	48.6	49.1	49.6	49.9	50.2	50.5	53.0
60	28.4	38.4	45.1	48.0	49.7	50.8	51.7	52.3	52.8	53.3	53.6	54.2	54.6	55.0	55.3	55.5	58.0
65	33.2	43.4	50.3	53.2	54.8	56.0	56.8	57.4	58.0	58.4	58.7	59.3	59.7	60.1	60.4	60.6	63.0
70	38.3	48.7	55.6	58.4	60.1	61.2	62.0	62.7	63.2	63.6	63.9	64.5	64.9	65.2	65.5	65.8	68.1
75	43.7	54.2	61.0	63.9	65.5	66.6	67.4	68.0	68.5	68.9	69.2	69.7	70.1	70.5	70.7	71.0	73.2
80	49.5	60.0	66.7	69.5	71.0	72.1	72.8	73.4	73.9	74.2	74.6	75.1	75.5	75.8	76.0	76.2	78.3
85	55.6	66.1	72.6	75.3	76.7	77.7	78.4	79.0	79.4	79.8	80.0	80.5	80.9	81.2	81.4	81.6	83.5
90	62.3	72.7	78.9	81.4	82.7	83.6	84.3	84.7	85.1	85.4	85.7	86.1	86.4	86.7	86.9	87.1	88.7
95	69.9	80.0	85.8	88.0	89.2	89.9	90.5	90.9	91.2	91.5	91.7	92.0	92.3	92.5	92.6	92.8	94.0
100	79.0	89.1	94.4	96.2	97.1	97.7	98.1	98.3	98.5	98.7	98.8	99.0	99.1	99.2	99.3	99.4	99.8

TABLE 8. Number of animals required in experimental and control groups to obtain statistical significant differences (example).

	Experimental group I	Control group II	
-pregnant (slaughter on day 3) hypothesis	30%	25%	wanted significans $p < 0.05$
Required no. of animals when probability of success should be at least 50%	700	700	
-pregnant (35 days) hypothesis	50%	45%	$p < 0.05$
Required no. of animals when probability of success should be at least 50%	800	800	

SUMMARY

1) In an arbitrarily selected clinically normal cattle population between 50-60% of the cows may be expected to concieve after an insemination irrespectible of whether it is the 1st, 2nd, 3rd, 4th, or 5th suc-cessive inseminations (Fig. 1).

2) In any cattle population (dairy cattle), large or small, the reproduc-tive efficiency should be determined by at least four parameters which may vary considerably within herds 1) conception rate (CR) = effecti-veness of inseminations, 2) heat detection rate (HDR), 3) days from calving to aimed 1st ins. (C-ins.), and 4) deadline for insemination of cows post partum in days (Table 5).

3) Regular/irregular interservice intervals occur in a normal cow popu-lation after a rather variable pattern due to a variety of factors, and it is questionable whether service intervals may be used to ca-tegorize the reproductive efficiency of cows as normal versus ab-normal (Table 6).

4) Significant differences ($p < 0.05$) in embryonic loss rates between experimental and control groups of cattle would necessitate planned slaughter and estimation of fertilizing failures as well as embryo-nic losses and require at least 4 groups of animals. From a statisti-cal point of view it may be possible to calculate the number of ani-mals which should be required in each group to obtain significant differences (Tables 7 and 8).

234

5) To undertake such experiments would require large groups of animals
which under most conditions are beyond the existing economic possibilities and besides additional variable factors may arise e.g.
estimation of embryonic quality, seasonal effect, a.o. which may produce conflicting results.

REFERENCES

Ayalon, A. 1978. A review of embryonic mortality in cattle. J. Reprod.
Fertil., 54, 483-493.

Hill, J.R. Jr., Lamond, D.R., Hendricks, D.M., Dickey, J.F. and Niswender,
G.D. 1970. The effect of undernutrition on ovarian function and fertility in beef heifers. Biol. Reprod., 2, 78-84.

Hoppe, F. 1954. En kritisk vurdering af den bovine sterilitetsbehandlings-
effektivitet. (Critical evaluation of the efficiency of the treatment
of bovine sterility). Nord.Vet.-Med., 6, 78-84.

Hyttel, P. 1982. Omløberproblematik og brunstsynkronisering. (The repeat
breeder term and heat synchronization in dairy cattle). Ph.D. Thesis,
Copenhagen. Royal Veterinary and Agricultural University.

Koefoed-Johnsen, H.H., Moustgaard, J., and Højgaard Olsen, N. 1955. Om vitamin B_{12}'s betydning for søer og gyltes frugtbarhed. (The importance
of vitamin B_{12} for the fertility of sows and gilts). Dansk Maaneds-
skr. Dyrlæg., 63, 1-11.

Pedersen, Kurt Myrup. 1982. Faktorer af betydning for reproduktionsfor-
løbet i malkekvægsbesætninger. (Factors of importance for the reproductive efficiency in dairy herds). 14th Nordic Vet.Congr., Copenhagen, 314-317.

Rasbech, N.O. 1981. Husdyrenes reproduktion 3. Gynækologi 4. (Reproduction in farm animals 3. Gynaecology 4). Textbook, p. 36. Carl Fr.
Mortensen A/S, Copenhagen.

Tanabe, T.Y. and Casida, L.E. 1949. The nature of reproductive failures
of cows of low fertility. J. Dairy Sci., 32, 237-246.

NUTRITION AND EMBRYO LOSS IN FARM ANIMALS

J. J. Robinson

Rowett Research Institute,
Bucksburn,
Aberdeen AB2 9SB,
Scotland

ABSTRACT

The literature on the effects of plane of nutrition on embryo survival in sheep, cattle and pigs is reviewed and evidence presented that extremes of nutrition in early pregnancy are detrimental to the growth and survival of the embryos of all three species. Deficiencies of a wide range of specific nutrients have been implicated in poor reproductive performance but it is only for selenium and vitamin E that the effect appears to operate through a shift in embryo survival. For ruminants, extended periods of feeding on the oestrogenic forages, in particular red clover, and the goitrogenic cruciferous crops reduces fertility and for both there is evidence of an effect on embryo survival. Possible modes of action of nutrition on embryo loss are, in the case of over nutrition, heat stress and, in the case of undernutrition, asynchrony of the embryo with its uterine environment, a disturbance in the amino acid composition of the uterine fluid or a reduction in the availability of glucose. Partial embryo loss not only reduces litter size but in the case of the ewe can result in the birth of lambs that are smaller than would be expected for their litter size.

INTRODUCTION

Embryo loss in all farm animals is usually much higher than can be attributed to chromosomal abnormalities. For example approximately 35 per cent of the fertilized ova in dairy cows in Britain are degenerate by 25 days post mating yet the incidence of abnormal karyotypes is only about 3 per cent (Gayerie De Abreu, 1984). Similarly for sheep, total embryo loss is often three to four times the 6-8 per cent (Long and Williams, 1980) caused by chromosomal abnormalities and cracked zona pellucida. In the pig, total embryo mortality is 30-40 per cent (den Hartog and van Kempen, 1980) and about one third of this arises from fertilized eggs which, as a result of recessive genetic traits and polyspermy are abnormal (Flint, Saunders and Ziecik, 1982).

Much of the difference between total embryo loss and that arising from the removal of abnormal karotypes is usually presumed to be avoidable in that it is thought to be linked to nutritional, environmental and/or disease factors. Although poor nutrition is often implicated in embryo loss, this paper illustrates that in certain circumstances high-planes of nutrition are also detrimental to embryo survival.

PLANE OF NUTRITION EFFECTS

Sheep

The results of experiments on the effects of level of feeding post mating on embryo survival are summarised in Table 1. Although the experiments listed in Table 1 form only a small selection of the total number on the subject, they have been chosen to illustrate the range of nutritional regimes employed and the variable effects that these regimes have on embryo survival and growth. In addition, where possible information is given on the body condition of the ewes at mating, since this measure provides an index of their premating nutritional status. Careful scrutiny of the data in Table 1 reveals a number of interesting trends that generally find support in the wider literature base from which the data in Table 1 were drawn. For example it would appear that the embryos of young ewes and older ewes that are in poor condition at mating are most at risk from low-plane feeding. Also, extremes of nutrition appear to be detrimental to embryo survival. From this observation comes the recommendation to keep, at maintenance levels of feeding during the first month of pregnancy, those ewes that have achieved maximum ovulatory response by being in the correct body condition at mating (score 3.0). Although there is a paucity of data on the effects of nutrition on highly-prolific breeds, a higher incidence of loss could be anticipated on the basis that nutritionally-induced losses appear to be more prevalent in twin- than single-ovulating ewes (Edey, 1976).

An interesting aspect of the information given in Table 1 is that fasting ewes for up to three days in the first fortnight of pregnancy has little, if any, effect on embryo survival. On the other hand extended periods of low-plane feeding (0.25 x maintenance) can, in the absence of embryo mortality, retard embryo development, as determined by the size of the pharangeal lobe and limb buds on day 21 (Parr, Cumming and Clarke, 1982). Similarly, low-plane feeding caused a 13 per cent reduction in embryo weight on day 35 (Parr and Williams, 1982). These observations

TABLE 1   EXAMPLES OF THE EFFECTS OF PLANE OF NUTRITION IN EARLY PREGNANCY ON THE SURVIVAL AND GROWTH OF SHEEP EMBRYOS
(Modified from Robinson, 1983)

Reference	Breed of ewe	Age (years)	Body condition at mating [a]	Plane of nutrition (x maintenance, M)	Stage of gestation	Summary of main findings
El-Sheikh et al (1955)	Shropshire, Hampshire, Oxford	1	Medium	High 2.0 M	0-40 days	High-plane feeding reduced embryo survival.
Edey (1970)	Merino	4-7	Fat	0.5 M	0-37 days	Extended interval to repeat oestrus suggesting embryo mortality.
Gunn, Doney and Russel (1972)	Scottish Blackface	5-6	Condition scores of 1.5 and 3.0	0.4 M v. 2.5 M	0-26 days	Embryo mortality higher for ewes in lower body condition (score 1.5); suggestion that high-plane feeding reduced embryo survival.
Cumming (1972)	Perendale	4-5	Fat	0.2 M	7, 14 or 21 days on 0.2 M in first 21 days	Embryo survival decreased as period of restricted intake was extended.
Blockey, Cumming and Baxter (1974)	Merino	Mature	Not given	Fasting	Fasted for 3 days starting on day 1, 5, 8, 10 or 12	Inconclusive; reduction in pregnancy rate in single-but not twin-ovulators
MacKenzie and Edey (1975)	Merino Merino	1.5 (Pri-miparous) 3.5-5.5	Not given	0.3 M 0.15 M	0-14 days 0-14 days	Extended interval to repeat oestrus. This suggests embryo mortality after day 12.
Cumming et al (1975)	Merino and Border Leicester x Merino	5	Condition score 2.5	0.25, 1.0 and 2.0 M	2-16 days	Embryo survival highest on 1.0 M.
Gunn, Doney and Smith (1979)	South County Cheviot	6	Condition score 2.0	0.7 M v. 1.75 M	0-30 days	Embryo survival improved by high-plane feeding particularly in ewes underfed before mating.
Parr, Cumming and Clarke (1982)	Merino	Mature	Not given	0.25 M v. 1.0 M	0-21 days	Retarded embryo development.
Parr and Williams (1982)	Merino	Mature	2-3	0.3 M v. 2.3 M	0-35 days	Reduction of 13% in embryo size by low-plane feeding.

[a] Description of the procedures for the subjective assessment of body condition score given by Meat and Livestock Commission (1981)

may explain the apparent carry-over effect on the birthweight of single lambs from ewes given an energy intake equivalent to 0.5 x maintenance between days 7 and 37 of pregnancy (Edey, 1970).

Cattle

For the few studies which deal specifically with the effects of plane-of-nutrition on embryo survival in the cow the nutritional regimes have been applied prior to mating (see Table 2) and interpretation of the observations requires information on the effects of nutrition on fertilization rate. Apart from the data for the small number of heifers studied by Hill et al (1970), the results of the larger studies presented by Spitzer et al (1978) and Dunn (1980) indicate that daily liveweight losses of up to 0.5 kg/day in cows that are in good condition at calving have no detrimental effect on fertilization rate. Rather the lower pregnancy rates are the direct result of an increase in embryo loss. Unlike the sheep there are very few observations for the cow on the interacting effects of body condition and plane of nutrition on embryo loss. For British Friesian heifers growing at 0.34, 0.50 and 0.68 kg/day, Leaver (1977) obtained pregnancy rates to first service of 67, 69 and 65 per cent respectively. However when the body condition of the heifers was taken into consideration the increase in the level of nutrition was accompanied by increases and decreases of 12 percentage units for those in 'poor/moderate' and 'good/very good' condition respectively. This detrimental effect on fertility of high-plane feeding of animals that are in good condition has also been reported by Ducker (1984) in an experiment involving 100 first-lactation cows and is in agreement with the findings for sheep which indicate that it is probably due to both a lower incidence of behavioural oestrus (Rhind et al (1984) and fertilization (Lamond, 1970), though Rhind et al (1983) argue from the observations of Gunn, Doney and Smith (1979) on South Country Cheviot ewes that embryo mortality may be involved also.

In 1970 the lack of data for the cow on the effects of nutrition on embryo mortality as distinct from fertilization failure prompted Lamond to suggest that on the basis of cost, size of animal and ease of handling, the sheep would form a convenient model for testing how nutrition influenced reproduction. With the advent of non-surgical procedures for the recovery of bovine embryos and their re-insertion into recipients by simple flank incision or trans-cervically this is no longer the case

239

TABLE 2   THE EFFECTS OF PLANE OF NUTRITION IN EARLY PREGNANCY ON THE SURVIVAL OF COW EMBRYOS

Reference	Breed of cow	Age	Plane of nutrition (Expressed in terms of daily live weight change, kg)	Stage of gestation	Summary of main findings
Hill et al (1970)	Angus and Hereford Angus	1.5–1.7 year-old heifers	+0.57 v. –0.23	–15 to +3 days –15 to +18 days	For those gaining weight 5 out of 5 and 3 out of 5 had fertilised ova on days 3 and 18 respectively. Corresponding values for those losing weight were 3 out of 5 and 2 out of 5.
Spitzer et al (1978)	Angus	1.3–1.6 year-old heifers	+0.02 v. –0.49	–21 to +4 days	79% fertilization for heifers gaining and 88% for those losing weight. Suggests that poor fertility on low-plane feeding is due to embryo loss.
Dunn (1980)	Not given (suckled)	Not given (cows)	+1.08 v. –0.31	–25 to +35 days	24% embryo loss for those gaining and 41% for those losing weight.

240

(Sreenan, 1983) and it can only be a matter of time before accurate
quantitative data on the effects of nutrition on embryo loss in cattle are
available.

Pigs

In Table 3 the same general format that was used for sheep and
cattle is adopted to summarise the range of nutritional regimes and their
effects on embryo loss in the pig.   Unlike cattle and sheep, low planes
of nutrition which involve losses in body weight in early pregnancy are
not applicable to modern systems of pig production.   Nonetheless it is of
interest that losses of up to 0.4 kg/d in bodyweight appear to have no
detrimental effect on embryo survival (King and Young, 1957) nor indeed
does fasting for up to three days immediately after mating (Ray and
McCarty, 1965)   Turning now to planes of nutrition that either maintain
or cause an increase in bodyweight then the conclusion drawn from many of
the early studies is that high-plane feeding promotes an increase in both
the number of ovulations and the incidence of embryo loss.   In general
the increase in embryo loss exceeds the increase in ovulation rate with
the result that litter size is smaller for animals kept on a high compared
with a low plane of nutrition prior to and immediately after mating.   The
first observations to disagree with this general conclusion are those of
Heap et al (1967) in that they show no effect of daily gains within the
range 0.1 to 1.0 kg on embryo survival.   These observations differ in two
important ways from the earlier studies.   Firstly they were made on sows
and not maiden females (gilts) as in the earlier experiments and secondly
the treatments were not applied until after mating.   In view of the
finding from reciprocal embryo transfer studies that there are detrimental
effects of high-plane feeding in early pregnancy on embryo survival in the
gilt and that it is the uterine environment and not the quality of the
embryo that is at fault (Bazer et al 1968) the disparity between the early
studies and those of Heap et al (1967) would appear to be due to
differences in age of animal.   Support for this view comes from the
recent study of Toplis et al (1983) which shows that a high plane of
nutrition which promotes daily liveweight gains of 0.8 kg during the first
month of gestation has no detrimental effect on embryo survival in mature
sows.   One reason for adopting a high plane of nutrition in early
pregnancy would be to replace body tissue lost in the previous lactation.
In some circumstances this may be both necessary and beneficial if one is

TABLE 3   THE EFFECTS OF PLANE OF NUTRITION IN EARLY PREGNANCY ON THE SURVIVAL OF SOW EMBRYOS

Reference	Age of sow	Plane of nutrition (Expressed, where possible, in terms of daily live weight change, kg)	Stage of gestation	Summary of main findings
Robertson et al (1951)	Gilts	Full fed v. restricted	Treatments applied during rearing and up to 25 days of gestation	Restricted-fed gilts had a higher embryo survival than full fed.
Self et al (1955)	Gilts	0.35 v. 0.44 kg/d		Higher ovulation rate but also higher embryo mortality on the high-plane of nutrition.
King and Young (1957)	2nd, 3rd & 5th parity	-0.38 v. +0.10 kg/d	1-28 days	Higher barren rate in low-plane group but negligible differences in embryo mortality.
Gossett & Sorensen (1959)	Gilts	0.34 v. 0.39 kg/d	Treatments applied during rearing and up to 40 days of gestation	Improved embryo survival in the slower-growing groups. An average of 1.3 more ova produced by high group but 15% more live embryos in those on lower plane of nutrition.
Sorensen et al (1961)	Gilts	0.36 v. 0.41 kg/d		
Ray and McCarty (1965)	Gilts	Fasting	Fasting for 0, 1, 2 or 3 days post mating	No effect of duration of fast on embryo survival.
Heap et al (1967)	Sows	0.10 v. 0.97 v. 1.06 kg/d	0 to 28 days	No effect on embryo survival or on the weight and size of the embryos but weights of membranes declined with increasing plane of nutrition.
Frobish and Steele (1970)	3rd parity	12.5 v. 29.3 MJ of ME/d	Complete gestation	A smaller number of piglets born on the high-plane of nutrition.
Dyck and Strain (1980)	Gilts	0.14 v. 0.40 kg/d	0-30 and 0.35 days	Embryo survival 75.8 and 86.2% for high and low groups respectively. Average embryo mortality 3.0 and 1.8 for high and low groups respectively.
Toplis et al (1983)	6th and 7th parity	0.34 v. 0.84 kg/d	Days 3 to 30	No effect on embryo survival.

to avoid the detrimental effect on embryo survival of low-plane feeding in
the previous lactation suggested by Hughes et al (1984).   For
fast-growing gilts however it would appear that embryo survival can be
improved by restricting food intake to maintenance needs for the first 10
days of gestation and thereafter increasing it again (Dyck and Strain,
1980).

SPECIFIC DIETARY NUTRIENTS

Sheep

The literature on the effects of specific nutrients on reproduction
in the ewe was reviewed by Robinson (1983) with the conclusion that
correction of deficiencies of protein and of some of the trace elements,
in particular copper, manganese, zinc, iron, cobalt and selenium improves
lambing percentages.   It is only for selenium however that there is
evidence that at least part of the improvement comes from a reduction in
embryo mortality (Hartley, 1963; Piper et al 1980).

Cattle

The fertility of lactating cows consuming low-quality roughages
increased from 36 to 75%, presumably as a result of greater incidence of
ovarian activity, when they were given daily protein supplements in the
form of either 900 g of cotton or groundnut cake (Topps, 1977).   On the
other hand excess protein can lower fertility, probably as a result of
reduced embryo survival (Rattray, 1977).   Although dietary deficiencies
of Ca, P, Cu, Zn, Mn, Co, Se, I and vitamins A, C and E and β-carotene
have all been implicated in reduced fertility so too have dietary excesses
of Ca, P, Na, Mo and F.   For many of the nutrients it is not clear
whether their effect is to impair ovulation through a depression in the
activity of the hypothalamic-pituitary axis, to reduce fertilization rate
or to interfere with embryo survival through effects on the quality of the
embryo and/or the uterine environment.

Pigs

The early literature on the effects of dietary deficiencies of
protein, fat minerals and vitamins on reproduction in the pig was
summarised by Tassell (1967) and more recently Rattray (1977) has drawn
attention to responses in litter size to calcium, zinc, selenium, choline
and vitamin A, as has Simmins and Brooks (1983) for biotin and Matte et al
(1984) for folic acid, but it is not known if these operate through a

reduction in embryo mortality.   In practice the use of balanced rations
in modern systems of pig production ensures that deficiencies of specific
nutrients are unlikely to occur.

FEEDS THAT REDUCE EMBRYO SURVIVAL

The oestrogenic forages, in particular red clover, with its high
content of the isoflavone, formononetin, which is converted in the rumen
into the phyto-oestrogen, equol, or 7,4-dihydroxyisoflavan, is a
well-recognised cause of reduced fertility in sheep and cattle.   Although
it is generally regarded that the adverse effect of the phyto-oestrogen is
mediated through a reduction in fertilization there is evidence that
prolonged grazing of oestrogenic forages can lead to cystic endometrial
hyperplasia and reduced embryo survival during the late-implantation phase
(see Review by Robinson, 1983).   Other examples of specific feeds that
interfere with embryo survival during implantation are the cruciferous
crops but it is uncertain if the increased embryo mortality is due to a
goitrogenic effect, anaemia or a reduced copper status.   If the cause is
a general anaemia then the use of varieties low in the anaemia factor,
S-methyl-cysteine sulphoxide (Smith, 1974) should alleviate the problem.

THE MODE OF ACTION OF NUTRITION ON EMBRYO LOSS

At present one can only speculate on the possible mechanisms by which
nutrition affects embryo survival and growth.   To explain the detrimental
effects of high-plane feeding in early pregnancy on embryo survival in
gilts and in ewes and cows that are in good condition at mating it is
tempting to implicate heat stress for there is ample evidence that during
the early cleavage stages embryos are particularly vulnerable to small
increases in maternal temperature (Edey, 1976).   Wilmut and Sales (1982)
take the view that some embryos die because of a spontaneously-arising
asynchrony with their uterine environment but evidence for a modifying
effect of nutrition on asynchrony has yet to be established.

In view of the extremely small requirements of the embryo for
nutrients it is not surprising that fairly extreme nutritional regimes are
required to affect their survival and these may operate through an
alteration in the endocrine balance of the mother.   For example, in
general there is an inverse relationship between the concentrations of
progesterone in plasma and plane of nutrition (Williams and Cumming, 1982)

and in separate studies it has been shown that the growth of sheep embryos in the first two weeks of pregnancy is influenced by the progesterone concentrations in maternal plasma (Wintenberger-Torres, 1976; Lawson, 1977). There is also evidence that enhancing the circulating concentrations of plasma progesterone either directly by supplementation or indirectly by administration of hCG during the luteal phase, can improve the fertility of both ewes (Kittok et al 1983) and cows (Sreenan and Diskin, 1983). Alternatively, in the experiments of Parr and Williams (1982) the reduction in the growth of sheep embryos caused by maternal undernutrition was accompanied by marked reductions in maternal blood glucose concentrations. This observation prompted the authors to postulate that embryo growth, and presumably in some instances survival, may be controlled by a limitation in the supply of maternal glucose in response to nutritional stress. To support their hypothesis the authors draw on the observations of Ellington (1980) who found that rat embryos cultured in the serum of fasted rats exhibited retarded growth and development unless the glucose concentrations of the serum were raised to those in the serum of unfasted controls.

IMPLICATIONS OF NUTRITIONALLY-INDUCED EMBRYO LOSS

It is well recognised that losses of fertilized ova prior to the initiation of implantation manifest themselves in a high incidence of repeat oestrous cycles at the normal interval, and/or in the case of the ewe and sow a reduced litter size. In contrast nutritionally-induced embryo losses during the implantation phase have a wider range of effects. These include delayed returns to oestrus, and/or in the case of the ewe and sow a reduced litter size; also in the ewe the birth of smaller than average lambs for their litter size. The evidence for this latter effect comes from the results of Rhind et al (1980) and McDonald et al (1981) which show that for prolific ewes the death of embryos in the third and fourth weeks of pregnancy disturbs the natural balance in the distribution of the embryos between the two uterine horns. In doing so it increases the within-litter variation in subsequent foetal growth and, as a consequence of the inability of the surviving embryos to use the maternal cotyledons vacated by those embryos that die, it also reduces the potential growth rate of those that survive.

CONCLUSIONS

In the absence of an alternative explanation for poor reproductive performance in a dairy or beef herd, a ewe flock or a sow unit, poor nutrition, operating through the hidden-loss of embryo mortality, is often implicated. While the evidence presented in this review does not disagree with this view, it suggests that for animals in good body condition at mating, a fairly severe and extended period of undernutrition is required to cause significant reductions in the growth and survival of their embryos. On the other hand, for fast-growing heifers and gilts and for animals that are in very good condition at mating, high planes of nutrition in early pregnancy are also detrimental to embryo survival. Deficiencies of specific nutrients can cause a reduction in reproductive performance but it is only for selenium and vitamin E that there is evidence that the effect operates directly on embryo survival.

REFERENCES

Bazer, F. W., Clawson, A. J., Robison, O. W., Vincent, C. K. and Ulberg, L. C. 1968. Explanation for embryo death in gilts fed a high energy intake. J. Anim. Sci., 27, 1021-1026.

Blockey, M. A. de B., Cumming, I. A. and Baxter, R. W. 1974. The effect of short term fasting in ewes on early embryonic surivival. Proc. Aust. Soc. Anim. Prod., 10, 265-269.

Cumming, I. A. 1972. The effect of nutritional restriction on embryo survival during the first three weeks of pregnancy in Perendale ewes. Proc. Aust. Soc. Anim. Prod., 9, 199-203.

Cumming, I. A., Blockey, M. A. de B., Winifield, C. G., Parr, R. A. and Williams, A. H. 1975. A study of relationships of breed, time of mating, level of nutrition, liveweight, body condition and face cover to embryo survival in ewes. J. agric. Sci., Camb., 84, 559-565.

den Hartog, L. A., and van Kempen, G. J. M. 1980. Relation between nutrition and fertility in pigs. Netherlands J. Agr. Sci., 28, 211-227.

Ducker, M. J. 1984. Effect of nutrition and management practice on fertility. Proc. Joint B.V.A. and B.S.A.P. Conf. on "Dairy cow fertility". Bristol, pp. 68-80.

Dunn, T. G. 1980. Relationship of nutrition to successful embryo transplantation. Theriogenology 13, 27-39.

Dyck, G. W. and Strain, J. H. 1980. Post-mating feed consumption and reproductive performance in gilts. Can. J. Anim. Sci., 60, 1060.

Edey, T. N. 1970. Nutritional stress and preimplantation mortality in Merino sheep, 1967. J. agric. Sci., Camb., 74, 193-198.

Edey, T. N. 1976. Embryo mortality. In "Sheep Breeding" (Ed. G. J. Tomes, D. E. Robertson and R. J. Lightfoot). (New England University Press, Armidale) pp. 400-410.

Ellington, S. K. L. 1980. In vivo and in vitro studies on the effects of maternal fasting during embryonic organogenesis in the rat. J. Reprod. Fert., 60, 383-388.

El-Sheikh, A. S., Hulet, C. V., Pope, A. L. and Casida, L. E. 1955. The effect of level of feeding on the reproductive capacities of the ewe. J. Anim. Sci., 14, 919-929.

Flint, A. P. F., Saunders, P. T. K. and Ziecik, A. J. 1982. Blastocyst-endometrium interactions and their significance in embryonic mortality. In "Control of Pig Reproduction" (Ed. D. J. A. Cole, and G. R. Foxcroft). (Butterworths, London) pp. 253-275.

Frobish, L. T. and Steele, N. C. 1970. Influence of energy intake through three gestations on reproductive performance of sows. J. Anim. Sci., 31, 200.

Gayerie de Abreu, F. 1984. A cytogenetic investigation of early stage bovine embryos - relation with embryo mortality. Abstract No. 44. Winter Meeting of the Society for the Study of Fertility, London.

Gossett, J. W. and Sorensen, A. M. 1959. The effect of two levels of energy and seasons on reproduction phenomena in gilts. J. Anim. Sci. 18, 40-47.

Gunn, R. G., Doney, J. M. and Russel, A. J. F. 1972. Embryo mortality in Scottish Blackface ewes as influenced by body condition at mating and by post mating nutrition. J. agric. Sci., Camb., 79, 19-25.

Gunn, R. G., Doney, J. M. and Smith, W. F. 1979. Fertility in Cheviot ewes. 3. The effect of level of nutrition before and after mating on ovulation rate and early embryo mortality in South Country Cheviot ewes in moderate condition at mating. Anim. Prod., 29, 25-31.

Hartley, W. J. 1963. Selenium and ewe fertility. Proc. N.Z. Soc. Anim. Prod., 23, 20-27.

Heap, F. C., Lodge, G. A. and Lamming, G. E. 1967. The influence of plane of nutrition in early pregnancy on the survival and development of embryos in the sow. J. Reprod. Fer., 13, 269-279.

Hill, J. R., Lamond, D. R., Henricks, D. M., Dickey, J. F. and Niswender, G. D. 1970. The effects of undernutrition on ovarian function and fertility in beef heifers. Biol. Reprod., 2, 78-84.

Hughes, P. E., Henry, R. W. and Pickard, D. W. 1984. The effects of lactation food level on subsequent ovulation rate and early embryonic survival in the sow. Anim. Prod., 38, 527.

King, J. W. B. and Young, G. B. 1957. Maternal influences on litter size in pigs. J. agric. Sci., Camb., 48, 457-473.

Kittok, R. J., Stellflug, J. N. and Lowry, S. R. 1983. Enhanced progesterone and pregnancy rate after gonadotrophin administration in lactating ewes. J. Anim. Sci., 56, 652-655.

Lamond, D. R. 1970. The influence of undernutrition on reproduction in the cow. Anim. Breed Absts, 38, 359-372.

Lawson, R. A. S. 1977. Research application of embryo transfer in sheep and goats. In "Embryo Transfer in Farm Animals" (Ed. K. J. Betteridge). (Monograph 16. Canada Department of Agriculture) pp.72-78.

Leaver, J. D. 1977. Rearing of dairy cattle. 7. Effect of level of nutrition and body condition on the fertility of heifers. Anim. Prod., 25, 219-224.

Long, S. E. and Williams, C. V. 1980. Frequency of chromosomal abnormalities in early embryos of the domestic sheep (Ovis aries). J. Reprod. Fert., 58, 197-201.

MacKenzie, A. J. and Edey, T. N. 1975. Short-term undernutrition and prenatal mortality in young and mature Merino ewes. J. agric. Sci., Camb., 84, 113-117.

Matte, J. J., Girard, C. L. and Brisson, G. J. 1984. Folic acid and reproductive performance of sows. J. Anim. Sci., 59, 1020-1025.

247

Meat and Livestock Commission 1981. Feeding the Ewe. (Meat and
    Livestock Commission, Bletchley, England).
McDonald, I., Robinson, J. J. and Fraser, C. 1981. Studies on
    reproduction in prolific ewes. 7. Variability in the growth of
    individual foetuses in relation to intra-uterine factors. J. agric.
    Sci., Camb., 96, 187-194.
Parr, R. A., Cumming, I. A. and Clarke, I. J. 1982. Effects of maternal
    nutrition and plasma progesterone concentrations on survival and
    growth of the sheep embryo in early gestation. J. agric. Sci.,
    Camb., 98, 39-46.
Parr, R. A. and Williams, A. H. 1982. Nutrition of the ewe and embryo
    growth during early pregnancy. Aust. J. Biol. Sci., 35, 271-276.
Piper, L. R., Bindon, B. M., Wilkins, J. F., Cox, R. J., Curtis, Y. M. and
    Cheers, M. A. 1980. The effect of selenium treatment on the
    fertility of Merino sheep. Proc. Aust. Soc. Anim. Prod., 13.,
    241-244.
Rattray, P. V. 1977. Nutrition and reproductive efficiency. In
    "Reproduction in Domestic Animals" (Ed. H. H. Cole and P. T. Cupps).
    (Academic Press, N. York, San Francisco, London) pp. 553-575.
Ray, D. E. and McCarty, J. W. 1965. Effect of temporary fasting on
    reproduction in gilts. J. Anim. Sci., 24, 660-663.
Rhind, S. M., Gunn, R. G. and Doney, J. M. 1983. A note on reproductive
    performance and plasma progesterone level during early pregnancy of
    Scottish Blackface and Cheviot ewes in relation to body condition and
    level of nutrition prior to mating. Anim. Prod., 37, 455-458.
Rhind, S. M., Robinson, J. J. and McDonald, I. 1980. Relationships among
    uterine and placental facotrs in prolific ewes and their relevance to
    variations in foetal weight. Anim. Prod. 30, 115-124.
Robertson, G. L., Casida, L. E., Grummer, R. H. and Chapman, A. B. 1951.
    Some feeding and management factors affecting age at puberty and
    related phenomena in Chester White and Poland China gilts. J. Anim.
    Sci., 10, 841-866.
Robinson, J. J. 1983. Nutrition of the pregnant ewe. In "Sheep
    Production " (Ed. W. Haresign). (Butterworth, London) pp. 111-131.
Self, H. L., Grummer, R. H. and Casida, L. E. 1955. The effects of
    various sequences of fall and limited feeding on the reproductive
    phenomena in Chester White and Poland China Gilts. J. Anim. Sci.,
    14, 573-592.
Simmins, P. H. and Brooks, P. H. 1983. Supplementary biotin for sows:
    Effect on reproductive characteristics. Vet. Rec., 112, 425-429.
Smith, R. H. 1974. Kale poisoning. In "Annual Report of Studies in
    Animal Nutrition and Allied Sciences, 30, 112-131. (Rowett Research
    Institute, Aberdeen).
Sorensen, A. M., Thomas, W. B. and Gossett, J. W. 1961. A further study
    of the influence of level of energy intake and season on reproductive
    perfromance of gilts. J. Anim. Sci., 20, 347-349.
Spitzer, J. C., Niswender, G. D., Seidel, G. E. and Wiltbank, J. N. 1978.
    Fertilization and blood levels of progesterone and LH in beef heifers
    on a restricted energy diet. J. Anim. Sci., 46, 1071-1077.
Sreenan, J. M. 1983. Embryo transfer procedure and its use as a research
    technique. Vet. Rec., 112, 494-500.
Sreenan, J. M. and Diskin, M. G. 1983. Early embryonic mortality in the
    cow: Its relationship with progesterone concentration. Vet. Rec.,
    112, 517-521.

Tassell, R. 1967.   The effects of diet on reproduction in pigs, sheep and
     cattle.   2: Pigs - protein, fat, vitamins, minerals and other
     dietary substances.   Br. Vet. J., 123, 170-176.

Toplis, P., Ginesi, M. F. J. and Wrathall, A. E. 1983.   The influence of
     high food levels in early pregnancy on embryo survival in multiparous
     sows.   Anim. Prod., 37, 45-48.

Topps, J. H. 1977.   The relationship between reproduction and
     undernutrition in beef cattle.   Wld. Rev. Anim. Prod., 13, No. 2,
     43-49.

Williams, A. H. and Cumming, I. A. 1982.   Inverse relationship between
     concentration of progesterone and nutrition in ewes.   J. agric.
     Sci., Camb., 98, 517-522.

Wilmut, I. and Sales, D. I. 1982.   Does hormonal imbalance cause death of
     embryos in sheep?   Ann. Rpt. Animal Breeding Research Organisation,
     Edinburgh, pp. 25-30.

Winterberger-Torres, S. 1976.   Action de la progesterone et des steroides
     ovariens sur la segmentation des oeufs chez la brebis.   Ann. de
     Biol. Anim. Bioch. Biophys., 7, 391-406.

UTERINE PATHOGENS AND EMBRYONIC MORTALITY

R. Bouters

Department of Obstetrics and Reproduction,
State University, Ghent, Belgium.

ABSTRACT

In recent experimental work the etiologic position of non-specific uterine pathogens in embryonic mortality (EM) has been thoroughly questionned. In cattle only C. pyogenes infection has an unequivocal carry over effect on subsequent fertility. Most of the specific infections causing EM have been eradicated but Campylobacter f. is still more important than it has been presumed. The possible role of viruses in bovine EM deserves much more attention. In swine several viruses are found associated with the SMEDI syndrome. The transplacental infection has been established for enteroviruses, parvovirus and Aujeszky virus. In horses and sheep uterine infections obviously play a minor role in EM.

INTRODUCTION

The embryonic period is the period extending from conception to the completion of the stage of differentiation (Committee on bovine reproductive nomenclature, 1971). In cattle this covers approximately the first 45 days of gestation.

The impact of embryonic wastage on reproductive efficiency is widely recognised in all mammalian species (for review see Boyd, 1965 ; Vandeplassche, 1968) including the human female (Edmonds et al.1982). The relative contribution however of different factors is still heavily debated. Most of the etiologic factors associated with embryonic mortality (EM) such as fertilisation failure, genetics, nutrition, hormonal status, uterine environment and aging of the gametes have been discussed in previous papers. In one of the final sessions the relative contribution of uterine pathogens in EM will be analysed.

Running through the literature a striking evolution is obvious. In the earlier days EM was mostly identified with uterine infection. Albrechtsen (1910) found that of 4881 cows considered as infertile 3925 or 80.4% had to be treated because of endometritis (Table 1).

TABLE 1   Percentage of cows treated for infertility.

Criterion	number	percentage
Total number of cows	8988	100.0
Treated for sterility	4881	52.1
Total number treated	4881	100.0
- for endometritis after retention	995	20.4
- for endometritis without retention	2930	60.0
- for cystic ovaries	156	3.2
- other reasons	800	16.4

Until recently it has been accepted for granted that endometritis is an important cause of infertility. Up to 1971 it has been claimed that endometritis was present in 55% - 81% of infertile cows (Brus, 1954 ; De Bois and Van Den Akker, 1957 ; Dawson, 1963 ; Sagartz and Hardenbrook, 1974). The published evidence by Stegenga, (1950)that "enzootic sterility" in cattle was essentially a venereal disease caused by Vibrio fetus venerealis (recently named Campylobacter f. ven.) further substantiated this general belief. Still in 1965 Boyd stated in an otherwise excellent review that infections, both specific and non-specific, play an important role in infertility and abortion in all our domestic animals. Steadily however, a growing sceptism about the etiologic role of uterine pathogens in reduced fertility became obvious. And when Vandeplassche in 1957 wrote that endometritis was one of the main causes of "sterility sine materia", his statement at the VIth Int.Congress on Animal Reproduction and AI (1968) that the relation between non-specific uterine infections, EM and repeat breeding was far from being proven, clearly demonstrated the changing philosophy.

Nowadays a distinction is made between specific and non-specific uterine infections.

I. Non-specific uterine infections.

The literature up to 1977 on the role of non-specific

bacteria has been extensively reviewed by Hartigan (1977). He
reported a substantial and spontaneous reduction of uterine
infection from almost 100% infection in the first 3 weeks post
partum to approximately 30% during the fifth week. From that
period on, the incidence of infection remained fairly constant
but the bacteria isolated changed often because of spontaneous
clearance and reinfection. De Bois (1961) found that most
cows had completely eliminated the infection when examined at
the time of first insemination.

The Irish group reported furthermore an almost identical
infection rate (Table 2) and type of infection (Table 3) in
repeatedly sampled fertile and repeat-breeder cows.

TABLE 2   Infection rates in fertile and repeat-breeder cows
           in repeated samples of 63 dairy cows.

	Fertility status	
	Fertile	Repeat-breeder
No. of cows sampled	46	17
No. of cows with at least 1 infected sample	26 (56.5%)	10 (58.8%)
No. of uterine samples	116	70
No. of infected samples	33 (28.4%)	20 (28.6)

(Hartigan, 1978)

Further analysis of the data revealed that the presence of
a non-specific uterine infection or endometritis around the
time of service was not an important cause of infertility in
cows which were free of clinically detectable uterine disease
(Griffin et al., 1974). Only C. pyogenes was consistently associated with endometrial lesions. Their severity and their
effect on fertility were mainly determined by the duration of
the infection (Hartigan et al., 1974). Theoretically, the
bacteria may interfere with fertility by
1. Directly killing the gametes or conceptus
2. Changing the uterine "milk"
3. Causing endometritis (toxic products, luteolysis)
4. Producing chronic histologic lesions.

TABLE 3   The bacteria isolated from fertile and repeat-bree-
           der cows.

	Fertile cows	Repeat-breeder cows
No. of cows sampled	46	17
No. of infected samples	33	20

Organism	No. of positive isolates	
Streptococci, α-haemolytic	7	6
Streptococci, β-haemolytic	1	1
Streptococci, non haemolytic	16	10
Staphylococci, coagulase +	3	0
Staphylococci, coagulase -	8	5
Corynebacterium pyogenes	3	0
Klebsiella spp	0	2
Diphteroids	5	1
Entero bacteria	1	1
Bacillus spp	1	1
Proteus spp	0	1

(Hartigan, 1978)

Some of these possible mechanisms recently have been studied
using intrauterine diffusion chambers as a model system in
rabbits (Hartigan & Defalla, 1984). The preliminary results
indicated that neither the presence of diffusable bacterial
products nor a persistent active infection had any "carry over"
effect on fertility. However, induction of uterine leucocyto-
sis by installation of glycogen, a leucocyte chemostatic agent,
caused complete termination of pregnancy and significantly
reduced fertility before and during implantation in rats.
Later stages remained unaffected (Anderson & Alexander, 1979).
So although some progress has been made, many aspects are
still open for discussion. However, the disappointing effects
of postbreeding antibiotic installions as a routine treatment
for repeat-breeding (Dohoo, 1983) and of using a plastic
sheath over the artificial insemination instrument for the
prevention of bacterial contamination during insemination

(Richards et al., 1984) raise profound doubts about the role of the vaginal and uterine bacterial flora in embryonic mortality in cattle.

However, with steadily increasing herd size and more cows being kept under loose conditions or in cubicles, the transmission of Corynebacterium pyogenes infection inside the herd will become ever more important and rigorous hygienic measurements will have to be instituted in order to keep the infection load within acceptable limits (Bouters, 1983).

In swine too, the presence of an un-specific uterine flora is compatible with pregnancy but Scofield et al. (1973) found a significantly greater embryonic loss in pigs with infected uteri than in those with sterile uteri.

II. Specific uterine infections.
a. Cattle

Although a great number of organisms (bacteria, viruses, protozoa, fungi) are involved in enzootic abortion, only few of them will interfere with normal embryonic development.Only venereal Trichomoniasis and Campylobacteriosis, Brucellosis and IPV (-IBR) infection are unequivocally associated with increased EM rates. Brucellosis and Trichomoniasis are almost eradicated in the western hemisphere. The same has been presumed for venereal Campylobacter infections, especially since efficient curative and preventive measurements have been developed (Bouters et al., 1973 ; Clark et al., 1974). Recent screening tests however have revealed that still a fairly high infection rate is present in West Germany both in AI bulls (Bisping et al., 1981) and in natural service (Dedié et al., 1982). Experimental soundings in different European, American and African countries have given similar results (Bouters, unpublished).

Although in artificial breeding, the presence of venereal vibrosis will be greatly masked by the addition of antibiotics to the extender, the permanent presence of infected bulls in AI stations which originally have been founded as a campaign against venereal diseases, should be limited by all possible means.

In natural breeding the pathogenesis and symptoms will
largely depend upon the immunological status of the female population. In recent outbreaks with almost no immunity in the
female population, organisms pass the cervical canal and enter
the uterus within 5 days after infective copulation. In the
uterus, they can induce endometritis, early embryonic death,
late embryonic death or early abortion (10%).

So the main clinical symptom is a drastic drop in NR
rate with over 50% prolonged oestrous cycles. In 10% of the
animals even salpingitis and permanent sterility will be found
(Vandeplassche et al., 1963). In chronic infections the fertility drop will be less dramatic and mainly confined to the
heifer population because of a local immunity developing in
infected females. Bulls however remain carriers for the rest
of their lives.

Under conditions of natural breeding IPV infection (infectious pustular vulvovaginitis) remains restricted to the
vagina and an almost normal NR is noted even in the presence
of a purulent vaginal discharge. There have been a number of
reports on outbreaks of IPV infection in AI bulls. When semen
contaminated with IPV virus was used, pustular endometritis,
infertility and shortened oestrus cycles were observed (Kendrick & Mc Entee, 1967 ; Boeckx et al., 1968 ; White & Snowdon, 1973).

It has been established that several viral infections
can be transmitted by the semen during AI service even before
clinical signs do appear in the infected bulls (Sellers et
al., 1968). Viruses reported in bovine semen include those of
foot and mouth disease, blue tongue, bovine leukaemia infectious bovine rhinotracheitis, bovine viral diarrhea, ephemeral
fever, lumpy skin disease, bovine enteroviruses, paravaccinia
and several uncharacterized viruses (Kahrs et al.,1980). Only
IPV-IBR virus has been unequivocally associated with reduced
fertility but whether this is due to a direct embryotoxic action or via a pustular endometritis is not yet elucidated. Two
other viruses, blue-tongue and bovine viral diarrhea-mucosal
disease virus, have been suspected to reduce fertility after

intra-uterine deposition, but further clarification is required (Archbald et al., 1979). These investigations however are
hampered by technical problems such as the high toxicity of
semen in cell cultures and costs involved in individual testing
of semen specimens.

b. Swine

In pregnant sows a clinical entity known as SMEDI (Stillbirth, Mummification, Embryonic Death, Infertility) and originally described by Dunne (1965) has been extensively studied.
The Smedi-viruses are enteroviruses belonging to different
serogroups (A, B, C, D, E) and with an enzootic character.When
sows become infected before 35 days of gestation, embryonic
death - either partial or complete - followed by resorption
will occur. Infections at later stages will induce mummification. The best prevention is to get the sows infected and
immune before the reproductive age.

Subsequently other viruses have been found associated
with the Smedi-syndrome including Porcine parvo virus, Aujeszky virus,Hog Choleravirus and picorna (Mengeling, 1980).
Other infections as Leptospirosis and Chlamydiosis are rather
associated with abortion and mummification at a later stage of
pregnancy. There are no comprehensive data on the incidence
of these viral diseases in embryonic wastage but it may be assumed that porcine parvovirus is the main causal agent in embryonic and foetal death. The most common route for infection for prenatal pigs is transplacental and during acute infection virus is shed by the semen. The major and usually
only clinical symptom of infection is maternal reproductive
failure with increased returns to oestrus, smaller litters
and mummified foetuses depending upon the stage of pregnancy
at which infection occurs.

Transplacental infection has also been described for
Aujeszky virus. For Hog Cholera most embryonic deaths and
mummifications originate from vaccination with attenuated live
vaccines.

c. Other domestic animals.

In horses, contagious equine metritis (CEM) is a highly

contagious genital infection, spread venereally and was first described in 1977 by Crowhorst. CEM is caused by Haemophilus equigenitalis, a gram negative coccobacillus. The infection is characterized by endometritis and infertility. Mares usually return to oestrus after an abbreviated dioestrus period (Timoney, 1982). Although CEM made a spectacular appearance and international spread, the infection appears to regress again, probably because of an intensive curative and preventive campaign.

In sheep, ureaplasmosis (T mycoplasmosis) appears to be a venereal disease possibly capable of causing infertility either by killing the ovum or by interfering with normal foetal development (Livingston & Gauer, 1981). More experimental work however will be needed before definite conclusions can be drawn.

In conclusion it can be anticipated that non-specific infections are not a major cause of embryonic death. In cattle the main infectious agent is still Campylobacter fetus and also viruses might be involved in some EM of unknown origin. In swine, several viruses are found associated with EM and actively attact the embryo and foetus via the transplacental route. In other farm animals uterine infections play obviously a minor role in EM.

REFERENCES

Albrechtsen, K. 1910. Die Sterilität der Kühe, ihre Ursachen und ihre Behandlung unter Berücksichtigung des Seuchenhaften Scheidenkatarrhs und des Verkalbens. R. Schoetz, Berlin.2.Auflage 1920 bearbeit von W. Stödter.

Anderson, D.J. & Alexander, N.J. 1979. Induction of uterine leukocytosis and its effect on pregnancy in rats. Biol. Reprod., 21, 1143-1152.

Archbald, L.F., Fulton, R.W., Seger, C.L., Al-Bagdadi, F. & Godke, R.A. 1979. Effect of bovine viral diarrhea virus on preimplantation bovine embryos : a preliminary study. Theriogenology, 11, 81-89.

Bisping, W., Kirpal, G., Sonnenschein, B. 1981. Die Diagnose und Bekämpfung der Campylobacter fetus subsp.fetus Infektion beim Besamungsbullen. Tierärztl. Umschau, 36, 667-674.

Boeckx, M., Ibrahim, M. & Bouters, R. 1968. Invloed van een
   acute IPV-IBR infectie op de bevruchtingsresultaten van
   KI stieren. Vl. Diergeneesk. Tijdschr., 37, 177-188.
Bouters, R., De Keyser, J., Vandeplassche, M., Van Aert, A.,
   Brone, E. & Bonte, P. 1973. Vibrio fetus infection in
   bulls : curative and preventive vaccination. Brit. Vet.J.
   129, 52-57.
Bouters, R. 1983. Ursachen und Frühdiagnostik von Fruchtbar-
   keitsproblemen in Rinderherden. Arb. Deutsch.Landw.Ges.,
   176, 61-66.
Brus, D.H.J. 1954. Biopsia uteri. Thesis Utrecht.
Boyd, H. 1965. Embryonic death in cattle, sheep and pigs.
   Vet. Bull., 35, 251-166.
Clark, B.L., Dufty, J.H.,Monsborough,M.J. & Parsonson, I.M.
   1974. Immunisation against bovine vibriosis. Vaccination
   of bulls against infection with Campylobacter fetus subsp.
   venerealis. Aust. Vet. J., 50, 407-409.
Committee on Bovine Reproductive Nomenclature. 1971. Recom-
   mendations for standardising bovine reproductive terms.
   Cornell Vet., 61, 216-237.
Dawson, F.L.M. 1963. Uterine pathology and bovine infertility.
   J. Reprod. Fert., 5, 397-971.
De Bois, C.H.W. & Van den Akker, S. 1957. Enkele opmerkingen
   over fertiliteitsprognose bij het rund. Tijdschr. Diergeneesk., 82, 951-971.
De Bois, C.H.W. 1961. Endometritis en vruchtbaarheid bij het
   rund. Thesis Utrecht.
Dedié, K., Pohl, R., Romer, H. Wagenseil, F., Albrecht, E. &
   Hühnermund, G. 1982. Zur Verbreitung, Ermittlung und Be-
   kämpfung der Venerischen Campylobacteriose (Vibrosis ge-
   nitalis) beim Rind in Beständen mit Bullenhaltung. Tier-
   ärtzl. Umschau, 37, 80-86.
Dohoo, I.R. 1983. A retrospective evaluation of postbreeding
   infusions in dairy cattle. Can. J. Comp.Med., 48, 6-9.
Dunne, H.W., Gobble, J.L., Hokanson, J.F., Kradel, D.C. &
   Dubash, G.R. 1965. Porcine reproductive failure associated with a newly identified Smedi group of picorna
   virus. Am. J. Vet. Res., 26, 1284-1296.
Edmonds, D.K., Lindsay, K.S., Miller, J.F., Williamson, E.&
   Wood, P.G. 1982. Early embryonic mortality in women.
   Fert. & Steril., 38, 447-453.
Griffin, J.F.T., Hartigan, P.J., Nunn, W.R. 1974. Non-specific uterine infection and bovine fertility. II.Infection
   patterns and endometritis before and after service. Theriogenology, 1, 107-113.
Hartigan, P.J., Griffin, J.F.T. & Nunn, W.R. 1974. Some observations on Corynebacterium pyogenes infection of the
   bovine uterus. Theriogenology, 1, 153-167.
Hartigan, P.J. 1978. The role of non-specific uterine infection in the infertility of clinically normal repeat-
   breeder cows. Vet. Sci. Comm., 1, 307-321.
Hartigan, P.J., Defalla, E.A. 1984. An experimental model to
   study the effects of uterine infection on fertility :
   some preliminary results. Proc. XIIth Int. Cong.Buia-
   trics, 830-832.

Kahrs, R.F., Gibbs, E.P.J. & Larsen, R.E. 1980. The search for viruses in bovine semen, a review. Theriogenology, 14, 151-165.

Kendrick, J.W. & McEntee, K. 1967. The effect of artificial insemination with semen contaminated with IBR-IPV virus. Cornell Vet., 57, 3-11

Livington, C.W. & Gaver, B.B. 1981. Effect of venereal transmission of ovine ureaplasma on reproductive efficiency of ewes. Am. J. Vet. Res., 43.

Mengeling, W.L. 1980. Porcine parvovirus infection in:Diseases of swine. 5th ed. Iowa State Univ.Press. pp.832

Richards, M.W., Spitzer, J.C., Neuman, S.K. & Thompson, C.E. 1984. Bovine pregnancy and non return rates following artificial insemination using a covered sheat. Theriogenology, 21, 949-957.

Sagartz, J.W. & Hardenbrook, J.J. 1971. A clinical, bacteriologic and histologic survey of infertile cows. J. Am. Vet. Med. Ass., 158, 619-622.

Scofield, A.M., Clegg, F.G., Lamming, G.E. 1974. Embryonic mortality and uterine infection in the pig. J.Reprod. Fert., 36, 353-361.

Sellers, R.F., Burrow, R., Mann, J.A. & Dwae, P. 1968. Discovery of virus from bulls affected with footh and mouth disease. Vet. Rec., 83, 303.

Stegenga, T. 1950. Vibrio foetus en enzootische steriliteit. Thesis Utrecht, pp.142.

Timoney, P.J. 1982. Isolation of the contagious equine metritis organism from colts and fillies in the United Kingdom and Ireland. Vet. Rec., 111, 478-482.

Vandeplassche, M. 1957. Die symptomenlose Unfruchtbarkeit bei unseren Haustieren. Wien. tierärztl. Mschr., 44, 449-461.

Vandeplassche, M., Florent, A., Bouters, R., Huysman, A., Brone, E. & De Keyser, P. 1963. The pathogenesis, epidemiology and treatment of vibrio fetus infection in cattle. IWONL verslagen over navorsingen n°29.

Vandeplassche, M. 1968. La mortalité embryonaire et son diagnostic. Proc. VIth World Cong. Anim. Reprod. & A.I., Vol. 1 , 347-391.

White, M.B. & Snowdon, W.A. 1973. The breeding record of cows inseminated with a batch of semen contaminated with infectious bovine rhinotracheitis virus. Austr. Vet. J., 49, 501-506.

PORCINE EMBRYO AND FETAL LOSS RELATED TO INFECTION
E. C. Simos

Veterinary Institute of Infectious
and Parasitic Diseases - Ierà odos
75 Botanicos, Athens (301), Greece

ABSTRACT
Infections are an important cause of porcine embryo and
f.etal losses and also sows infertility in Greece. Control measures include mainly,diagnosis, treatment and prevention. The
increased knowledge of the infectious aetiology of swine reproductive problems will contribute to the decrease of prenatal
and perinatal mortality which represents a serious economic
problem for pig farmers.

INTRODUCTION
Since 1981, when Greece joined the EEC as a full member,
problems have arisen in certain sectors of livestock production
and especially in the pig industry as far as adapting to the
new situation is concerned (Kyriakis, 1982).
The main reasons for these problems have been a) a rapid increase in the consumption of products of animal origin and,
b) the low productivity of all farm animals, pigs included.
Nevertheless, over the last 15 years, pigmeat production has
expanded rapidly and now supplies more than 25% of the country's total requirement for meat (Katsaounis,1980; Kyriakis,
1982).

In general, the productivity of the Greek pig industry is
low. The main problems which account for this low productivity
are scouring problems of piglets, respiratory diseases,conta-
gious diseases (swine fever, erysipelas) and problems related
to embryo and fetal losses and infertility in sows. Data on
the occurence of abortions and stillbirths are not available,
but it is believed that these account for approximately one-
fourth of the losses suffered in pigs of preweaning age.

In our paper we will be dealing with terms as abortion,
stillbirth,fetal death, embryonic death, mascerated fetuses
and mummified fetuses.

BIBLIOGRAPHICAL REVIEW

The etiology of the porcine embryo and fetal losses remain undetermined in many instances.

The infectious etiologic factors known to be associated with abortion, stillbirth and infectious infertility are listed below (Rasbech, 1969; Dunne, 1970; Scofield et al, 1974; Kirkbride and McAdaragh, 1978; Suzuki, 1980).

Protozoa

Toxoplasma gondii

Bacteria

Brucella suis

Leptospira sp.

Escherichia coli, paracolon organisms

Streptococcus sp.

Staphylococcus aureus

Pasteurella multocida, P. *ureae*

Erysipelothrix insidiosa

Corynebacterium pyogenes

Salmonella sp.

Pseudomonas aeruginosa

Listeria monocytogenes

Bacillus sp.

Chromobacterium sp.

Mycobacterium tuberculosis

Mycoplasma

Fungi

Aspergillus fumigatus

Nocardia asteroides

Fusarium gramineatum

Viruses

Swine fever

Aujesky´s Disease (Pseudorabies)

Japanese encephalitis

Japanese hemagglutinating virus

English hemaglutinating virus (PPV)

SMEDI viruses

Foot-and-mouth disease
Swine influenza
Vesicular stomatitis
Transmissible gastroenteritis

Miscellaneous
Mastitis-Metritis-Agalactia (MMA)

GENERAL ASSESSMENT OF THE SITUATION IN GREECE
Bacterial Diseases

Brucellosis in pregnant sows is characterized by abortion
or the birth of weak pigs. The detection of the disease is
undertaken by serological testing. Aborting sows usually develop detectable titers. The antigen used in the standard agglu-
tination test is prepared from Brucella suis Strain 6.
The 56° C Heat Inhibition Test, C.F.T. and Rosebengal Test are
also used. The incidence of the positive results is low (1%).
During 1981-83, forty eight embryos were examined and four were found positive for *Brucella suis* (2) and *Brucella sp*. The
policy test and slaughter is used for brucellosis (Simos, 1971;
Greek Min.of Agr., 1983).

Several species of Leptospira are capable of infecting
swine. Recently seven cases of chronic leptospirosis were
found in modern industrial pig farms of S.Greece, with a high
level of Abortion as the main symptom. The causes were *Leptospira pomona*, *L.tarassovi* and *L.grippotyphosa*. Mummified embryos, mascerated fetuses and stillborn pigs were found (Chondou et al, 1984). In two serological surveys in N.Greece positive reactions were given by the following serotypes *L.icte-
rohaemorrhagiae brysawa*, *L.luisiana*, *L.canicola*, *L.pomona*, *L.
grippotyphosa*, *L.sejroe*, *L.tarassovi* and *L.panama* (Sarris et
al, 1984; Ekateriniadou et al, 1984).

Escherichia coli is recognised as a cause of abortion in
sows. Hemolytic or non-hemolytic E. coli strains have been
isolated from a number of fetuses and vaginal discharge involving cases of abortions, stillbirth and infertility. Isolations
were made from the stomach and other organs of the fetuses
which showed gastroenteritis and congestion of the viscera

(Frangopoulos and Simos, 1972).

In our laboratiry a hemolytic *Streptococcus sp.*was isolated from the viscera of two fetuses in two separate cases of porcine abortion. It appears that E. coli and streptococci are more than simple contaminants.

Staphylococcus sp. was isolated from various organs of fetuses more often than streptococci. From the history it was known that abortion occured between 2-3 months of pregnancy and the placenta was expelled normally. A purulent vaginal discharge noted in some cases was controlled by antibiotics (Simos , 1969).

The outbreaks of swine erysipelas in Greece create a serious economic problem, especially in the small units. Abortions,usually due to this septicaemic disease were reported during the febrile stage. *Erysipelothrix insidiosa* organisms have been isolated from aborted fetuses in two cases. Penicillin controls the disease effectively.

Organisms of the genus *Corynebacterium* are isolated often from aborted fetuses and vaginal discharge of the aborting sows. A hemolytic strain showing the biochemical tests of C.*pyogenes* was isolated from the organs of one fetus.

Pseudomonas aeruginosa was also isolated in our laboratory both from fetuses and vaginal discharge (Simos, 1974).

Pasteurella and Salmonella have never been isolated from the reproductive system of sows in any Greek laboratory, although infections due to these organisms are common for respiratory and gastrointestinal systems,respectively. The same is valid for mycoplasmas, a common isolant from respiratory and genital systems.

Fungal swine abortion has never been reported in Greece.

Viral Diseases
Epizootiological observations implicated the virus of swine fever (SF)as a cause of embryonic absorption, fetal death,malformation and stillbirth. The danger for the porcine embryo and fetus from the use of attenuated SF vaccines is nowdays wellknown , although we personally had stressed this earlier (Simos, 1968).

Experimental evidence has shown that the virus of SF invades the uterus of the susceptible pregnant sow at any stage of gestation causing embryonic death, absorption, fetus mummification and stillbirth. Congenital tremors and cerebrellar hypoplasia and hypomyelinogenesis also have been associated with prenatal infection with SF. Other lesions include hemorrhages of the skin and local necrosis of the liver (Tsirojiannis et al, 1969).

Today, we use the attenuated SF vaccine strain C very carefully only in danger areas.

There is strong evidence that Aujeszky Disease(Pseudorabies) causes death of embryos and fetuses, mummification, abortion between 1.5 - 2 months and reduces litter size. Many in our country believe that Aujezsky Disease was introduced to Greece recently with the importation of breeding animals from abroad.

Infection of swine takes place in the respiratory and digestive systems by inhalation or ingestion. Sows and boars can also be infected by genital tract. Approximately 50% of the pregnant sows abort their fetuses 10-20 days after the onset of the clinical disease or around the 3rd month of pregnancy. Paired serum samples show increased titres during convalescence (Koumbati-Artopiou, 1978).

Lately the disease has been causing growing concern in Greece and now is a matter of anxiety to pig producers. Special measures are soon to be taken.

Sera collected from a small number of pigs of various ages in a large breeding herd have shown antibodies to 59 e/63 english hemaglutinating virus-porcine parvovirus (PPV, titres 1:640 - 1:16000). The unit had a history of abortions and stillbirth and infertility in sows. It is believed that these disorders are due to the existance of PPV. An extensive survey into this subject would show its role in these reproductive abnormalities and losses (Kyriakis et al, 1978).

Outbreaks of transmissible gastroenteritis (TGE) are rarely found in our units. Strong evidence of a clinical nature offers support to the belief that TGE is capable of causing a disturbance of embryonic growth if sows are infected during the first month of pregnancy.

In the past(before 1975) during outbreaks of foot-and-mouth disease, we had also observed abortions among other manifestations.

Swine vesicular disease was observed in Greece in 1979. Except lameness and ulcerations in the feet, no other clinical signs or lesions were found (Dimitriadis et al, 1979).

Miscellaneous

Cases of mastitis-metritis-agalactia (MMA) syndrome are quite common in all units. Bacteriological examination of the infected material reveals cocci and E. *coli* (Simos,1969).

REFERENCES

Chondou,A., Kyriakis,S., Simos,E., Stoforos, E., Xylouri,E.
and Giannacopoulou, L. (1983). Chronic Leptospirosis in
pigs. Cases of sow abortion in industrial farms. Bull.
Hell.Vet.Med.Soc. 34, 318-325.

Dimitriadis,J., Pappous,C. and Brovas, D. (1979). A case of
swine vesicular disease in Greece. Bull.Hellenic Vet.Med.
Soc. 30, 265-276,

Dunne, H.W (1970). Abortion, Stillbirth, Fetal Death and Infectious Infertility. Diseases of Swine, p.836-850. The
Iowa S.Un.Press, Ames, USA.

Ekateriniadou,L., Sarris,C. and Papadopoulos,O. (1984). Serological survey for leptospirosis in swine farms of Thessaloniki area. Proc. 3rd Greek Veterinary Congress, Corfu,
p.112.

Frangopoulos,A. and Simos, E.(1972). E. coli infections in
animals diagnosed in Veterinary Institute of Microbiology.
Proc. 5th Nat. Congress on Bacteriology, Athens, p.284-292

Greek Min.of Agr. (1983). Data on activities of Veterinary Service. Athens.

Katsaounis,N.(1980). Pig Industry. University of Thessaloniki
Press.

Kirkbride ,C.A. and J.P.Mc Adaragh (1978). Infectious agents
associated with fetal and early neonantal death and abortion in swine. JAVMA, 172, 480-483.

Koumbati-Artopiou, M.(1978). A comparative study of serological
methods for the diagnosis of Aujeszky's disease in experimentally and naturally infected swine. Thesis.Vet.Col-
lege Univ. of Thessaloniki.

Kyriakis,S.. Andreotis, J., Papasolomontos, S., Henry, R.R.
and Tsaltas,C. (1978). Research on the existence of hae-
magglutination inhibiting antibodies to porcine parvovirus in pig sera in Greece.Bull. Hell. Vet. Med. Soc.,
29, 235-238.

Kyriakis,S., (1982). Pig production in Greece. Pig news and
information, 3, 419-421.

Rasbech,N.O. (1969). A review of the cases of reproductive
    failure in swine. The Brit.Vet. J. <u>125</u>, 599-616.
Sarris,K., Saoulidis,K., Ekateriniadou,K., Mazaraki,K. and
    Burtzi,E. (1984). Leptospirosis in a pig farm in the area
    of Thessaloniki. 3rd Greek Veterinary Congress, Corfu,
    p. 110.
Scofield, A.M., Clegg, F.G. and Lamming G.E. (1974). Embryonic
    mortality and uterine infection in the pig. J.Reprod.
    Fert., 36, 353-361.
Simos, E. (1968). Contribution to the study of post-vaccinal
    reactions and losses in swine immunized with modified live
    virus hog cholera vaccines.Bull. Hellenic Vet.Med.Soc.,
    <u>19</u> ,111-121.
Simos, E. (1969). Toxaemic agalactia syndrome in sow. A report
    of two cases. Veterinary News-Greece, <u>2</u>, 11-14.
Simos, E. (1971). Results of a serological survey for porcine
    brucellosis in Northern Greece. Proc. 4th Nat. Congress
    on Bacteriology, Thessaloniki p. 149-155.
Simos, E. (1974). Pseudomonas infection in animals and birds.
    Proc. 4th National Congress on Bacteriology, Athens,
    p. 58-69.
Tsirogiannis, E., Leontides, S. and Exarchopoulos, G. (1969).
    Hog cholera. The centarl nervous system lesions and their
    diagnostic significance in relation to the typical gross
    lesions. Veterinary News <u>1</u>, 3-15.
Suzuki, T. (1980). Fetal death in Swine with the isolation
    of Pasteurella ureae. J.Japan. Vet.Med.Ass. <u>33</u>, 219-222.

DISCUSSION

Courot:

Is there any data on the effect of inadvertently immunising pregnant ewes?

Boland:

No.

Wilmut:

Is there an effect of treatment on embryo size, birth weight or gestation length?

Boland:

There is no evidence of an effect on birth weight. I do not have data on embryo size or gestation length.

Koch:

When and how were antibody titres established?

Boland:

On Day 7 after the booster injection blood samples were assayed.

Diskin:

Is there any comparative data on the effect of immunisation on ovulation rate and litter size?

Boland:

There are several reports indicating increased ovulation rate and litter size in breeds such as Border Leicester crosses, Merino, Corriedale and Romney.

Wilmut:

Do antibodies to other steroids produce the same breed effects?

268

Boland:

We have not looked at embryonic mortality following immunisation against other steroids.

Heap:

Whereas sheep embryos produce little if any oestrogen during the preimplantation period, they do metabolise other steroids extensively e.g. androstenedione and progesterone. The effect you have demonstrated on embryo survival may be related to an interaction between anti-androstenedione antibodies and endogenous steroids in the early embryo.

Sreenan:

You suggest a 25-30% abnormality rate of PMSG induced oocytes. The data of Gayerie (1982) suggests that only 3-5% of PMSG induced superovulated ova have chromosomal abnormality. Can you reconcile these figures?

Greve:

The proportion of abnormality matured oocytes exhibiting abnormal meiotic figures is similar to that reported by Gayerie (1982) for abnormal ova. A further proportion will either fail to become fertilised or will be lost as embryonic deaths at a later stage.

Fischer:

Are there other than morphological methods for assessing embryo quality and is it possible to predict the subsequent viability of an embryo at the 2-8 cell stage?

Greve:

In domestic animal embryo transfer programmes morphological assessment is currently the only practical method. The use of FDA or DAPI fluorescent dyes do not add to morphological assessment. The best method of assessing 2-8 cell embryos would be in vitro culture.

Wilmut:
 Should more emphasis be placed on embryo-uterine synchrony?   Should exogenous progesterone be given to synchronise the recipient uterine environment with the requirement of the embryo?

Sreenan:
 This would not be practical.   The variation in endogenous progesterone levels and the difficulty of increasing these by exogenous progesterone would not allow very precise synchronisation of embryo and uterine environment.

Niemann:
 What is the embryonic loss rate after the transfer of frozen-thawed embryos?

Greve:
 The pregnancy rate following transfer of thawed embryos of good morhpological appearance is about 5-7% lower than for similar quality fresh embryos.

Bolet:
 Is the ranking of males for embryonic mortality the same for either fresh or frozen semen?

Courot:
 I am not aware of documentation on this point.

Diskin:
 There is Scandanavian literature indicating a sire effect on embryonic mortality which arises in the daughters of the sire.   Is there similar evidence for sheep?

Courot:
 There is some evidence for an indirect effect of the male on fertilisation and embryonic mortality rates of their daughters.

Fischer:

Does frequency of mating increase the proportion of
immature sperm and in turn the embryonic mortality rate?

Courot:

High frequency of ejaculation does not accelerate
epididymal transport of spermatazoa from the head and
corpus epididymis, it only does so for mature sperm present
in the cauda epididymis. High frequency of mating may
reduce fertilisation rate because of the reduction in sperm
numbers but will not increase embryonic mortality.

Niemann:

How does the male influence embryonic loss after
fertilisation?

Courot:

The answer can only be speculative. During
spermiogensis and epididymal maturation many changes occur
in the chromatin structure and organisation related to DNA-
protein relationships and the formation of disulphide bridges
etc. It is possible that spermatazoa with an incomplete
chromatin structure may be capable of fertilising ova which in
turn may develop but only for a short period.

Bouters:

Is there a male age effect on fertilisation?

Courot:

Probably yes. There is a known male age effect on
fertilisation rate and an established correlation between
fertilisation rate and embryonic survival.

Fischer:

Is triploidy or polyploidy lethal for early embryos?

Mannaerts:

In most cases polyploidy leads to early embryonic

death.   There are, however, some reports that triploid
embryos have survived until birth or shortly after birth.

Bouters:
     The repeatability of Feulgen staining is very low.
Can you please comment on this?

Mannaerts:
     The Feulgen staining measurements were carried out in
collaboration with the departments of histochemistry and
cytochemistry at Leiden.   Co-workers there have
successfully standarised a Feulgen staining methodology
which may also be used for bovine embryo nuclei.

Niemann:
     What cytogenetic method did you use for the preparation
of nuclei and metaphases and what percentage of embryos
yielded usuable metaphase figures?

Mannaerts:
     The method used was that of King et al. (1970) and is
similar to the air-drying method for chromosome preparations
described by Tarkowski (1966).   The percentage of usuable
metaphase spreads depended on whether vinblastine or colchicine
was used.   Embryos cultured with vinblastine were all sexed
while embryos cultured without vinblastine or colchicine were
sexed in 50 percent of cases.   Metaphase spreads suitable for
sex determination were also suitable for the detection of
polyploidy or centric fusion translocation.

Bolet:
     The statistical models you described are interesting and
models should be developed for polytocous as well as monotocous
species.

Rasbech:
     So far we have only addressed the question of dairy
cattle but it will be necessary to develop models for sheep

and pigs.

Sreenan:

What is the earliest stage at which you have observed
the effect of embryo death on the birth weight of the
surviving foetuses?

Robinson:

Certainly by Day 25 but I think perhaps earlier. We
have quantitative data on the effect of foetal position and
distribution in utero on foetal size. Distribution and the
initiation of implantation starts as early as Day 13-15 in
sheep and, therefore, placental size and foetal growth could
be affected from this time.

Heap:

Your data suggests a local inhibition of placental
development arising from early embryonic death in ewes
carrying more than one embryo and leading to low birth
weights of singletons. Carunclectomy studies reducing
placentome number suggest, however, that compensatory
hyperthrophy of the placenta occurs. Would you comment on
this?

Robinson:

There is some compensatory hypertrophy of the placenta
but in our studies this has not been sufficient to support the
rate of foetal growth we observe where there is no evidence
of embryonic mortality.

van der Lende:

Is it possible that an unfavourable uterine environment
might have an effect on embryo survival rate and also on the
growth of the surviving embryo (s)?

Robinson:

Yes.

Bouters:

In twin-pregnancy in the mare one foetus usually occupies one entire horn and uterine body while the second is confined to the top of the other horn. This reduction in uterine space causes a high level of embryonic mortality because of the reduced nutrition supply to the confined foetus.

Wilmut:

Is there published data on the uterine temperature of animals on a high plane of nutrition soon after mating?

Robinson:

No but high environmental temperatures which result in a small increase (about $1^{\circ}C$) in maternal temperature induce high levels of embryonic death during the first few days after mating. I am speculating when I suggest that high plane feeding could through problems in the dissapation of heat lead to an increase in body temperature, for example in heavily fleeced ewes.

Sreenan:

Your review suggests that only extreme nutrition levels affect embryo mortality, yet most of embryonic loss we are discussing occurs under moderate to good management conditions.

Robinson:

Yes, I agree but that is the evidence from the literature. We may have to be careful with high plane feeding in early pregnancy, particularly in warm environments as I have already suggested that embryo mortality may increase from heat stress in animals on a high plane of nutrition.

Rasbech:

Would you comment on the diagnosis of Vibrio infection in bulls?

Bouters:

The "heifer test" is an excellent diagnosis. However, the combination of improved culture techniques, millipore filtration and immunofluoresence results in a 98% recovery of organisms from preputial washings of infected bulls.

Diskin:

Is B.V.D. implicated in embryonic mortality?

Bouters:

There is no evidence to suggest this.

Rasbech:

Parvovirus vaccine is effective and now available. Gilts should be treated one month before service. Have you any experience of this?

Simos:

The vaccines are not yet available in Greece and extensive research will be necessary before recommending their widespread use.

GENERAL CONCLUSIONS AND OUTLOOK

R.B. Heap

AFRC Institute of Animal Physiology,
Babraham, Cambridge CB2 4AT, U.K.

During the last two days we have heard an excellent series of papers which confirm that embryonic mortality remains a serious problem so far as efficient animal production is concerned. Clearly the problem is a complex one involving physiological, endocrinological, genetic and immunological components, and it is frequently difficult to identify which one or more of these possible causes is most relevant in modern farming practice. The question of whether embryonic mortality is greater in conditions of intensification and high productivity remains controversial and during the course of the seminar few speakers directed attention exclusively to this point. It is important to recall, however, that pioneer studies of Brambell in the 1940's showed a high incidence of preimplantation loss (up to 13%) in wild rabbits in certain areas, with a total loss of up to 43% so the problem is not simply one of domestication. During the course of this seminar, several speakers expressed concern that after a period of about 50 years during which much has been written and documented about the prevalence of early embryo losses in farm animals, little can be said about recommended therapy apart from prophylactic measures designed to ameliorate the effects of high ambient temperature and to avoid stressful stimuli during early gestation.

Much research has been done to specify the degree of synchronisation required between the embryo and it's environment to achieve a successful pregnancy. The hormonal requirements of early pregnancy have also been well defined. Yet studies designed to supplement deficiencies of hormone secretion, or to identify associations between fertility and factors such as nutrition have frequently proved ambiguous or even contradictory. In fact it has often proved difficult to deduce anything about the causal factors of early embryonic

mortality from such studies except in certain instances where
infectious agents or heat-induced loss have been implicated.
A critical analysis of these aspects shows that we remain
ignorant of the major factors that precipitate early embryo
loss in farm animals, and there are very good reasons for this.

Identifying the reasons for our lack of knowledge may
help to define research needs, and one impression that emerges
from our discussions is that the continuation of past
strategies is unlikely to provide further insight into the
problem, but rather more of the same type of information.
Among immediate needs is the development of a simple method
that allows detection of the time when an embryo fails, and
an approach to this problem using early pregnancy factor (EPF)
has been described. The detection of EPF remains cumbersome
and inconsistent so that the search for better methods must be
a high priority. We lack good models in which for example,
perturbation of the bi-directional signalling would allow us
to distinguish between factors that are obligatory and those
that are solely permissive for the establishment of pregnancy.
Immunisation against androstenedione in Merino sheep seems to
cause retardation of embryo development and interferes with
implantation, and selective ablation using immunological
means may provide an important tool for the recognition of
specific and indispensible signals. We lack knowledge of
why fertilization with ram spermatozoa that are prevented from
normal epididymal maturation results in increased embryonic
mortality and arrested cleavage and delayed development, but
the results are reminiscent of recent experiments in mice
which highlight the significant contribution of the male
pronucleus to preimplantation embryonic development. While
we are unable to explain adequately why asynchrony between
the embryo and it's uterine environment gives a low success
rate for pregnancy in farm animals, new information about the
production of uterine factors such as small polypeptide growth
factors and secreted proteins that may regulate embryo growth
suggests leads for future research.

There are those who advocate the further use of large
scale surveys to establish correlations by statistical analysis,

but others claim that the effects of certain hormones on embryo survival are only detectable in longitudinal studies in which each animals acts as it's own control, as seen in the inverse relationship between basal progesterone levels in circulation shortly after fertilization and subsequent embryo loss in sheep.   Few will doubt, however, that the problem of embryonic mortality is ripe for new scientific initiatives if progress is to be made in reducing the wastage that occurs in terms of animal productivity and welfare.   In this respect priorities that hold promise for the future include clarification of how and when embryonic genes are first expressed and to what extent expression is influenced by the maternal environment; how the transcription of genes that code for products aimed at prolonging luteal function are regulated in the early embryo; the nature of mechanisms by which the maternal immune system is modified to facilitate acceptance of the fetal allograft; and further elucidation of the genetic factors involved in the basal embryo loss.

We are indebted to Dr. J. Sreenan for arranging and inviting us to this Symposium and to Dr. J. Connell for making it possible to hold a meeting on this topic at such an opportune time.

278

LIST OF PARTICIPANTS
****

. Beghelli, V.   Instituto di Fisiologica
                                  Veterinaria, Universita di
                                  Perugia, Via S. Constanzo, 4,
                                  06100-Perugia, ITALY

. Boland, M.   Faculty of Agriculture,
                                  University College Dublin,
                                  Lyons Estate, Newcastle,
                                  Co. Dublin, IRELAND

. Bolet, G.   INRA, Stn. de Genetique,
                                  Quantitative et Applique,
                                  78350 Jouy-en-Josas,
                                  FRANCE

. Bouters, R.   Faculty Vet. Medicine,
                                  State University of Ghent,
                                  Casinoplein, 24,
                                  B-9000 Ghent, BELGIUM

. Courot, M.   INRA,
                                  Stn. Physiologie Reproduction,
                                  Nouzilly, 37380 - Monnaie,
                                  FRANCE

. De Boer, P.   Department of Genetics,
                                  Gen. Foulkensweg, 53,
                                  Wageningen,
                                  NETHERLANDS

. DeLouis, C.   INRA,
                                  Stn. Centrale de Physiologie
                                  Animale,
                                  78350, Jouy-en-Josas,
                                  FRANCE

. Connell, J.   Commission of the European
                                  Communities, Co-ordination
                                  of Agricultural Research,
                                  Rue de la Loi, 200,
                                  1049, Brussels, BELGIUM

. Diskin, M.G.   Agricultural Institute,
                                  Belclare, Tuam,
                                  Co. Galway, IRELAND

.   Fischer, B.   Abteilung Anatomie und
                                    Reproduktionsbiologie,
                                    Klininkum der RWTH Aachen,
                                    D-5100 Aachen,
                                    FEDRAL REPUBLIC OF GERMANY

.   Greve, T.   Royal Vet. & Agricultural
                                    University,
                                    Copenhagen 1870,
                                    DENMARK

.   Heap, R.B.   AFRC,
                                    Institute of Animal
                                    Physiology,
                                    Babraham,
                                    Cambridge CB2 4AT,
                                    UNITED KINGDOM

.   Helmond, F.   Dept. of Animal Physiology,
                                    University of Wageningen,
                                    Haarweg 10,
                                    Wageningen,
                                    NETHERLANDS

.   Koch, E.   Institute fur Tierzucht und
                                    Tierverhalter,
                                    Mariensee, D-3057 Newstadt,
                                    FEDRAL REPUBLIC OF GERMANY

.   Mannaerts, B.M.J.L.   Department of Functional
                                    Morphology,
                                    Faculty of Veterinary
                                    Sciences,
                                    State University,
                                    Yalelaan 1,
                                    3508 TD Utrecht,
                                    NETHERLANDS

.   Niemannm, H.   Institut fur Tierzucht und
                                    Tierverhalter,
                                    Mariensee,
                                    D-3057 Newstadt,
                                    FEDRAL REPUBLIC OF GERMANY

.   Parmigiani, E.   Universita Degli Studi di
                                    Parma,
                                    Facolta di Medicina
                                    Veterinaria,
                                    Instituto di Clinica
                                    Chirurgica Vieterinaria,
                                    Via de Taglio,
                                    Parma,   ITALY

. Rasbech N.O.   Institute for Animal
                                    Reproduction,
                                    Roual Vet. & Agric. Univ.,
                                    Copenhagen, 1870 V,
                                    DENMARK

. Robinson, J.J.   Rowett Research Institute,
                                    Bucksburn,
                                    Aberdeen,
                                    SCOTLAND, U.K.

. Simos, E.C.   Veterinary Institute of
                                    Infectious and Parasitic
                                    Diseases,
                                    IERA ODOS, 75, Botanicos,
                                    Athens, 301,
                                    GREECE

. Sreenan, J.M.   Agricultural Institute,
                                    Belclare, Tuam,
                                    Co. Galway, IRELAND

. van der Lende, Ir. T.   Department of Animal
                                    Husbandry,
                                    Agricultural University of
                                    Wageningen, Marijkeweg, 40,
                                    6709 PG Wageningen,
                                    NETHERLANDS

. Wilmut, I.   ABRO,
                                    Drydenfield Laboratory,
                                    Roslin, Midlothian,
                                    EH25 9PS,
                                    SCOTLAND, U.K.